Growth of a Prehistoric Time Scale

Based on Organic Evolution

Growth of a Prehistoric Time Scale

Based on Organic Evolution

WILLIAM B. N. BERRY
University of California, Berkeley

 W. H. FREEMAN AND COMPANY
San Francisco and London

Copyright ©1968 by W. H. Freeman and Company.

No part of this book may be reproduced by any mechanical, photographic, or electronic process, or in the form of a phonographic recording, nor may it be stored in a retrieval system, transmitted, or otherwise copied for public or private use without the written permission of the publisher.

Printed in the United States of America.

Library of Congress Catalog Card Number: 68-14224.

Preface

Determining the relative position of two or more events in time is of daily concern to everyone; and all of us, therefore, keep a set of units with which to relate events in the past to one another and to the present almost constantly in mind. The most familiar method is to note how long it takes a fixed point or line on the earth's surface to rotate past the sun. The measure of one such rotation is called a "day." Days may be grouped into longer units—weeks and years—or divided into shorter ones—hours, minutes, and seconds.

Before man began recording his history in terms of these units, the earth's surface underwent change, and organisms arrived on and left the stage provided by that surface. The history of these events during the long eons between formation of the earth and man's first recorded history is of primary concern to geologists. The time span during which these events took place is commonly termed "geologic" or "prehistoric" time.

Just as astronomers found a method for dividing recorded time based on the rotation of the earth, so earth historians have found methods for dividing prehistoric time. This book tells of the search that led to the development of a method for dividing prehistoric time based on the evolutionary development of organisms whose fossil record has been left in the rocks of the earth's crust. That search resulted in the growth and development of various units that have been linked together to form a loosely defined time scale. Many of the terms given to the units in that scale mean something different to almost every geologist. What one geologist means by "Silurian" may be the Silurian and Ordovician to a second geologist, and the Silurian, Ordovician, and Cambrian to a third; still another geologist may use "Gothlandian" in place of "Silurian." American geologists don't use the Carboniferous unit discussed in this book; and European geologists don't use the American Mississippian and Pennsylvanian units. In similar manner, other units discussed here have different meanings for different geologists. The scale of units is still growing and developing; this book is really an interim report.

Despite opinions (see Kitts, 1966) that the record of organic evolution through natural selection is of little value for the establishment of time units, it is in fact evolution that has provided a set of events that are unique in time by which the relationship of one event to others may be determined; that is, by which the passage of time may be reckoned. Evolution through natural selection comprises a set of continually active processes in which events that are unique occur. These events may be used as marker points in time so that a time unit may be established as the time span from one such marker to another.

As some of the principles upon which the time scale and its units may be based have their beginnings in man's early scientific inquiry, a history of the growth and development of a prehistoric time scale must begin among the earliest records of inquiry that may be considered scientific. The story follows the quest for the principles that were used to interpret the fossil record for its time significance and make the evolutionary events in that record the basis for time units. After the principles were established, various units of the time scale could be determined.

The first principles proposed for use in working with the evolution of organisms to determine time units led only to vaguely defined intervals. These were the periods and epochs recognized during the early part of the nineteenth century: units of relatively long duration, some apparently longer than others. By the middle of the nineteenth century the need had grown for units of lesser duration. A principle was developed that led to recognition of shorter time units that could be clearly delimited.

Our story is the growth of the units—both those of relatively long and those of relatively short duration. The story is not a closed book: work still goes on toward making all the units of the time scale clearly defined and, indeed, toward making a scale that will be a standard of reference for all geologists, a scale in which each of the units will have the same meaning to all.

Vast sequences of rock with few obvious fossils lie beneath the rocks containing abundant and obvious fossils. These sequences must, when division is based upon evolutionary development of organisms, be lumped into one unit, the Precambrian. Study of the radioactive decay of certain elements in some minerals of the Precambrian rocks indicates that this time span encompasses nearly 80 percent of the time from the formation of the earth's crust to the present. The time intervals based upon the evolutionary development of organisms thus cover only the last 20 percent of that time. Although the story presented here is concerned with only a short part of the time that has elapsed since the earth's crust formed, its several aspects have occupied and continue to occupy the thoughts and research interests of scores of geologists. The units of the time scale based on organic evolution today form the basic time framework of historical geology.

Although some of this story could have been written many years ago, it was not, possibly because most geologists seem to be more interested in the inorganic than the organic aspects of their science. Geologists have been quick to recognize the value of applying the principles of chemistry and physics in geologic research, but they have, with few exceptions, remained unaware of or uninterested in those principles of biology that have application in geology. Obviously, any geologic use of organismal remains should employ those biological principles that pertain.

Not all the time units ever proposed are mentioned in this book, for it is not an exhaustive catalogue of these units. Only the principles that led to recognition of the units and the most widely cited units (these are indicated in Table 1, page 9) are discussed. A more thorough listing and discussion of the era, period, and epoch names, with particular reference to the United States, is available in Bulletin 769 of the U.S. Geological Survey, compiled by M. Grace Wilmarth. Many more details of the history of paleontology and stratigraphy with reference to the periods has been given by Zittel (1899) in his renowned book on the history of paleontology, *Geschichte der Geologie und Paläeontologie*.

The criticisms of Professor R. M. Kleinpell of the University of California, Berkeley, have been extremely valuable at every stage in the preparation of the manuscript. I am also indebted to Professor S. W. Muller of Stanford University for many comments and suggestions and for reading an early version.

William B. N. Berry

April 1968

Contents

ONE   A Need for a Time Scale   1
Principles   2
The Economic Motive and a Need for a Time Scale   3

TWO   First Principles   11
Uniformity in Natural Processes   11
Superposition   23

THREE   Attempts at Division   26

FOUR   Darwin's Contribution—Organic Evolution Through Natural Selection   41

FIVE   A Maelstrom of Opinions   50
Faunal (and Floral) Succession   50
The Setting   60
The Transition from Descriptive to Interpretive Units   61

SIX   Growth and Development of the Periods   64
The Tertiary   64
The Carboniferous   66

The Cretaceous 69
The Jurassic 73
The Quaternary 76
The Triassic 78
The Silurian and the Cambrian 80
The Devonian 88
The Permian 91
The Ordovician 95
The Pennsylvanian and the Mississippian 99

SEVEN Lyell's Percentages 103

EIGHT An Overview of the Earth's Fossil Record 115
The Eras 116
The Precambrian 118

NINE Units with Boundaries 121
D'Orbigny's Stages 122
Oppel's Principle 126
Some Modern Modifications of Oppel's Method 129
Zone Names 133
"Zone"—Other Uses 134
Miocene Zones and Stages in California 134

TEN Marker Points and the Correlation Web 137

Literature Cited 147

Index 153

CHAPTER ONE A Need for
a Time Scale

Wonder, awe, fear, delight—these are some of the emotions that man's dwelling place has aroused in him since he first began to be aware of his natural surroundings. The earliest surviving records, most of which seem to be derived from oral tradition, indicate that he felt that although the earth had been created especially for him it was under the control of a whimsical deity whose actions were unpredictable. Then, as man became aware that certain events repeated themselves, and that by using the certainty of the repetition, he could predict other events, he concluded that most observable processes obeyed principles, or laws. It was in the sixth century B.C. that the Greeks began to record and analyze data from which the principles that govern nature were induced. This was the beginning of science in the Western World.

Unfortunately, after a start in the five or six centuries before Christ, recorded knowledge and that part of it commonly labeled science did not become enriched during the Middle Ages. Scientific thought was limited until well into the sixteenth century —with the exception of the activities of a few scholars among whom Roger Bacon (1210–1294) and Leonardo da Vinci (1452–

1519) may be noted for their diverse scientific as well as humanistic interests. Soon after, however, a host of scientific thinkers, including Copernicus, Descartes, Kepler, Galileo, and Steno, stimulated advances in science.

Many individual fields of science, one of which is geology—the science of the earth—grew from the inquiries into nature begun during the Renaissance. Geology is founded upon a number of principles, and a basic goal of geologists is the induction of such natural principles from among the masses of data that continually accumulate from observations of natural phenomena. After principles have been established, they may be used for further interpretation of earth's phenomena. A geologist's task is the assembly of a picture of the history of the earth and the phenomena, both biological and physical, that pertain to it. Because those principles (interpretive generalizations) of geology pertaining to fossils have long been confused with observations and descriptive generalizations, a brief review of what the scientist means by "principle" may be of help in keeping the descriptive aspects of stratigraphic geology separate from the interpretive aspects. A scientist tries to keep those terms and classifications appropriate to observable phenomena separate from those required for interpretation based upon principles.

Principles

Principles are the essence of science, and are indispensable to it. A principle is a well-tested interpretive generalization relevant to a highly consistent relationship between certain specific facts induced from many observations and descriptive generalizations. It is a general truth derived from and based upon observational data. An interpretive generalization must stand the test of repeated experiment and observation—the test of experience—to ascertain its validity. Then and only then is it a principle, and thereby available for deducing the generally true significance of particular relevant facts as they are encountered.

No principle can ever be considered absolutely and totally infallible: the best are excellent approximations. As powers of observation and experimentation are increased to permit greater refinement in study, a more widely applicable principle may replace one that has been valid for a long time. Principles thus pertain to the level of refinement of the observations upon which they are based. New equipment that permits more detailed and precise observations may lead to new principles, or to modifications of existing principles.

The Economic Motive and a Need for a Time Scale

A need for a time scale by which the time of formation of a rock unit might be related to that of others arose from mining activities and a general interest in minerals. During the sixteenth and seventeenth centuries, this interest spread to an increasing study of rocks themselves. Mining had been practiced for centuries, and various minerals and mining methods were so well known before the time of Christ that Diodorus Siculus, a Greek historian writing about 60 B.C., was able to point out the various mining methods used in several countries and to trace the trade routes of gold and silver and of some other metals as well. The Romans used metals extensively and seized metal-bearing deposits of communities they subjugated: they took over tin mines from the early inhabitants of Cornwall in the British Isles and silver deposits from the Carthaginians. The actual study of minerals received considerable stimulus from the writings of Georg Bauer, better known under his Latinized name, Agricola. He was a physician in a mining town in Germany, and wrote in the first half of the 1500's. His works include a sketch of the geographic distribution of various economically useful metals, an account of the growth of mining in Germany and Austria, and the classification and description of many minerals known in his time. Agricola is often called "the father of mineralogy" because of his perceptive classification scheme and mineral descriptions. He worked and lived in an area that was one of the mining centers of Europe, and his knowledge of and enthusiasm for minerals stemmed directly from his observations. During his life and from his time on, more and more attention was directed to the study of rocks for their potential economic value, and many institutions of learning established teaching positions in mineralogy. The economic motive was a real stimulus to studies leading to increased knowledge of the history of the earth.

As more mineral-rich areas were found and knowledge of them broadened, local rock relationships for individual areas were recognized by miners. Informal terms were commonly applied to particular rock units, and they sufficed to permit description and communication concerning any individual deposit. If one wished, however, to relate the rock succession in one mining area to the successions in other areas, he couldn't, for no general succession had been recognized; and no set of time units had been established by which the rocks in one area could be related by time of formation to those in other areas. The first attempt to allow definition of the relationships among several rock sequences of different areas was a very generalized succession. Even the general rock succession could not be recognized over a large area, and hence a need arose for something more than a purely descriptive grouping by smell (as in "stinking vein") or by gross aspect (as in "peacock vein" or "clunch"). Something in the rocks had to be interpreted to determine age relationships among rocks of widely separated areas.

Slowly fossils became valued as data that could be obtained from a large portion of the rocks of the earth's crust, and could

A Need for a Time Scale

be interpreted to ascertain time relationships. Charles Lapworth (1879, p. 3) made this point as follows:

> We have no reliable chronological scale in geology but such as is afforded by the relative magnitude of zoological change—in other words, that the geological duration and importance of any system is in strict proportion to the comparative magnitude and distinctness of its collective fauna.

W. P. Schimper (1874, p. 680) stated in this regard:

> . . . in spite of the evident continuity in the evolution of the organic kingdom through the geologic ages there can nevertheless be distinguished in this progressive and continuous movement a constant change in the grouping and relative development of types, a change which enables us to identify for each epoch, and even for each geologic period, a group of forms constituting what we call the organic character of the epoch or period.

As both Lapworth and Schimper noted, organic remains found in stratigraphic succession could be seen to change in kind. The changes were, at first, best seen in large aggregates of fossils collected in stratigraphic order from large groups of strata over a wide area. These large aggregates that were distinctive became the bases for time units. The time units founded upon large fossil aggregates were vague and poorly delimited. The time intervals represented by them were realized to be relatively long. Shorter intervals of time were delimited after detailed analyses of ancestor-descendant relationships of groups of species were carried out. Slowly a hierarchy of time units was developed. Each of the time units was established upon a unique fossil aggregate. The smallest of such time diagnostic aggregates have been termed congregations (Berry, 1964; 1966).

The time units were founded upon certain unique fossil aggregates found in rocks in particular areas in which the fossiliferous rock was well-exposed or in which its superpositional relationships could be demonstrated, or both. Such areas have been termed *type areas*. The rocks and the fossils they contain in the type areas form the observational foundation upon which the interpretations were built. In some analyses (Charles Lyell's Miocene, for example), supplemental type areas, in which the diagnostic fossils were more abundant or in which the superpositional relationships of the fossil-bearing rocks could be seen more easily than in the original type area, were used.

The type areas are important to geologists because the fossils that formed the data for the original interpretation were found there and others could be collected there. Indeed, the type areas permit the original "experiment" by which the time unit was recognized to be "re-run" as many times as geologists would care to. A geologist could collect fossils from the type area, re-study the stratigraphic superpositional relationships of the rocks bearing them, and then, hopefully, come to the same conclusions as did the original investigator in setting the particular fossil aggregate apart as the basis for a time unit.

Historically, two sets of hierarchical divisions have been used by geologists in the time scale based on the succession of fossils seen in the rocks of the earth's crust. One set of divisions is designated "time," and the other set is designated "time-stratigraphic." The time units have been said to be dependent upon the time-stratigraphic units because they refer to an abstraction, time, whereas a time-stratigraphic unit is the actual body of fossiliferous rock from which the fossils interpreted as being indicative of a time interval were collected. Both the time and its corresponding time-stratigraphic units have been given the same name because of the dependence of the time units on the time-stratigraphic. The terms commonly used for the time and time-stratigraphic hierarchical divisions are:

	Time Divisions	Time-stratigraphic Divisions
Era		—
	Period	System
	Epoch	Series
	Age	Stage
	Phase	Zone

These terms are not necessarily accepted by all geologists. The agreement that does exist is the result of many conferences carried on over many years. The conferences began to debate the matter of terms following a suggestion made by a group of American geologists at the First International Geological Congress held in Paris in 1878. Both international and national committees were appointed to examine the terminology then in use. A report given at the Second Congress in 1881 suggested use of most of the terms just listed. They became more widely used after the Eighth Congress in Paris in 1900, where the committee in charge of the debate over the terms recommended use of these nine terms. The use of the designation "time-stratigraphic" came from the recommendation made by two American geologists, S. W. Schenck and H. G. Muller, who in 1941 tried to clarify the distinction between the interpretive nature of the "time" and "time-stratigraphic" units in contrast with the purely descriptive rock or stratigraphic term "formation."

Understood by many geologists during the debate but misunderstood by many since that time is the nature of the term "formation." As Schenck and Muller emphasized, it is a descriptive-observational entity quite separate from the interpretive time and time-stratigraphic units. A formation is a cartographically useful rock unit. The use of the term for such a unit dates back to 1822 when W. D. Conybeare and W. Phillips used "formation" to mean a sequence of similar beds that had accumulated under similar environmental conditions and had a definite position in a succession of such units.

It should be pointed out that before the writing of Conybeare and Phillips, the word "formation" as used by geologists had a wide variety of meanings. To Abraham Werner, all rocks of one kind were a formation. All sandstones, in his view, constituted a formation distinct from all limestones. To other German geologists, a formation was a widely recognizable set of rocks that might comprise many local rock strata. They used the word in the same sense that French geologists of the late 1700's and early 1800's used "*terrain*." Following debate on the use of terms in the geologic time scale, it was realized that the German geologist's "*Formation*" and the French *terrain* meant the same as "system" in the recommended set of terms. In the English-speaking world at least, the use of system has replaced the use of both the German *Formation* and the French *terrain* for the time-stratigraphic unit.

Schenck and Muller (1941) also suggested using "group" to mean a group of formations. "Series" has been widely used to mean a sort of super group of formations: such a descriptive use of the word must be kept distinct from any time-stratigraphic use.

In much current work with strata and the fossils contained in them, it is convenient to use a time-stratigraphic unit as a general term to mean the rocks in a given area that can be demonstrated to have been deposited during a particular time interval. To demonstrate that certain rocks in an area were deposited during a particular phase, for example, fossils must be collected from the rocks, analyzed, and judged to be closely similar (in terms of the association of species and also in terms of the stage of evolutionary development represented by the species) to those species considered diagnostic of that phase. A time-stratigraphic zone includes the rocks deposited during a phase. The phase is defined as the time span from the appearances of particular new species in a stratigraphic section at a particular selected place to the appearances of certain other new species, which denote the beginning of the next phase. The time units period and epoch as discussed here were recognized following analysis of fossil aggregates. That organic evolution in some form (either through natural selection, special creation, or inheritance of acquired characteristics) lay behind the changes seen in the succession of fossil aggregates was not realized by geologists at the time the units were established. That evolution through natural selection is, in fact, the reason why

TABLE 1. Correlation between the time scale based upon organic evolution and that based on rotation of the earth.

Era	Period	Epoch	Approximate number of years ago (in millions)
CENOZOIC	Quaternary	Recent	
		Pleistocene	
			3
	Tertiary	Pliocene	12
		Miocene	25
		Oligocene	37
		Eocene	56
		Paleocene	
			65
MESOZOIC	Cretaceous		
	Jurassic		150
	Triassic		200
PALEOZOIC	Permian		250
	Carboniferous — Pennsylvanian		300
	Carboniferous — Mississippian		
	Devonian		350
	Silurian		
	Ordovician		450
	Cambrian		500
			600
PRECAMBRIAN			5000

there is a succession of fossil aggregates has not been widely understood since.

Further, it has not been widely realized that a time unit should mean an interval of time extending from events (or an event) that are unique in time and are used to denote its beginning (these events are included in the interval) to events (or an event) used to denote the beginning of the next time interval. Because time intervals should be so delimited, the periods and epochs based upon large fossil aggregates are only vague, poorly defined time units. They have not yet developed into clear-cut, discrete time intervals. When the ages and phases were established following the methods laid down by Albert Oppel in the mid-1850's, some precision did enter the growth of the time scale: His methods for analyzing fossils permitted recognition of those boundaries that are today considered phase boundaries.

In working with fossils, the events used to denote the beginning of a time interval may be speciations or other events in the evolution of life. When the expression of such an event is seen in the rocks, that particular expression at that particular position may be selected as a marker point at the beginning of a time interval. A net or web of correlations must be built out from that marker point.

The growth of a time scale based upon organic evolution through natural selection will be traced in the following pages. As is true of most growths, it has proceeded at a variable rate, and comprehension of the growth has also been widely variant from generation to generation of geologists. As the time scale is still being shaped and refined, this story does not concern a dead issue, but a still-continuing process.

CHAPTER TWO **First Principles**

Uniformity in Natural Processes

Before a time scale could be devised, and, indeed, even before geology could be a science, some principles had to be established. As inquiry into natural phenomena grew, the initial observations led to descriptive generalizations. Not until James Hutton published in 1795 the conclusions from his extensive examinations of rock relationships and natural processes at work upon the earth and Charles Lyell's subsequent amplification of Hutton's conclusions in his textbook *Principles of Geology* (published 1830–1833) did geology become a science founded upon principle. Until the documentation of the uniformity of nature's processes in Hutton's and Lyell's books had been given to scientists, geology foundered. Natural processes were not considered to be uniform throughout time in their actions, but were thought to be controlled by Providence. The earth's history and the processes active in nature were simply concluded to be the work of a provident Creator and to have neither great antiquity nor any remarkable invariability. Many naturalists of the early 1800's thought various catastrophes had wracked the earth in the past at the whim of the Creator, and that all things, both animate and inanimate, had been placed by him upon the earth following the last great catastrophe. Many considered the Noachian Deluge to have been that catastrophe, and placed blind faith in Bishop Ussher's pronouncement that 4004 B.C. had been the earth's birthdate.

Given this framework of thought concerning the history of the earth and natural phenomena, little wonder that most theories of earth development and natural phenomena were edifices built of utter speculation. The dogma propounded by Abraham Werner and Georges Cuvier, as we shall see, was based largely upon preconceived notions about the phenomena of nature with which all observations had to be made to fit. Hutton and Lyell placed conclusions based upon patient and thorough study of natural phenomena against the temples built of speculation. It is they who made geology a science. Hutton and, following him, Lyell, induced the basic principle upon which geology is founded.

At the heart of Hutton's conclusions concerning natural phenomena was the principle that, in general, natural processes and functions observable today have been going on in the same basic manner throughout past time. Both Hutton and Lyell emphasized that to interpret the significances of past phenomena, we depend upon what we can learn from those natural processes that we see operating with consistency or from those observable natural relationships that hold true consistently.

When Lyell wrote his *Principles of Geology,* he had at least two goals. One was to present an alternative to the catastrophic concept of earth history advanced by Cuvier and widely proclaimed by many naturalists of the early nineteenth century, of whom William Buckland was perhaps the most articulate. Lyell's second goal was firmly to position geology among the sciences by documenting that it is, as Hutton had demonstrated, founded upon principles. As S. J. Gould (1965) has pointed out, Lyell thus fought on two fronts to combat the idea that the earth and all things on it were the result of divine creation.

To refute the catastrophic concept of earth history, Lyell suggested that natural processes acted at relatively uniform or constant rates. To oppose the divine creation idea, Lyell stressed that natural laws, or principles, governed the processes active on and in the earth, and that these were, in general, invariable in space and time. Lyell emphasized that which Hutton had made clear in his work: that by studying present natural processes and relationships, those active and existing in the past might be understood.

Gould discussed Lyell's notion of uniformity in rate of change as "substantive uniformitarianism." He pointed out that it is not a demonstrable principle relevant to the interpretation of natural processes in the past. Uniformity in rate of change may also be termed "gradualism." It has, unfortunately, been widely discussed as a basic principle in geology. It has, as Gould noted, been criticized because of its inapplicability in interpretation of past phenomena.

Gould's "substantive uniformitarianism" has been confused commonly with Lyell's second assertion—that natural laws were, in general, invariable. That assertion is commonly discussed as the "principle of uniformitarianism." Gould termed it "methodological uniformitarianism" in an effort to clear up the confusion that had arisen over what is meant by the term "uniformitarianism." It should be noted here that uniformitarianism, where used in this work, has that meaning given to "methodological uniformitarianism" by Gould.

Because geology is founded upon observations of natural relationships that hold consistently true and upon use of these observations in interpretation of past phenomena, the history of man's awareness that such observations could be put to such a use will be traced. Perhaps the earliest recorded observations and conclusions were those of the Hellenes. These Greek scholars, more than 500 years before the birth of Christ, were confident that the natural phenomena of the world in which they lived could be understood, and they actively sought basic principles, or natural laws, from their observations. They saw evidence that changes in the positions of land and sea were effected by natural, observable processes.

One of the earliest Greek students of nature of whom we have a record was Xenophanes of Colophon who lived approximately 600 years before the birth of Christ. He observed—on mountaintops that were a great distance inland—shells not unlike those of clams then living along the shoreline. He also noted similar evidence that the sea had once covered a part of the land forming the island of Malta. He concluded from his observations that the land had been subjected to periodic sea incursions. Later, Xanthus the Lydian, an Ionian colonial, and Herodotus the traveler, both of whom lived just prior to 400 B.C., described shells entombed in rocks far inland that were similar to those of living clams and snails. They concluded that the ocean had once extended to the places in which those organic remains in the rocks had been found, and that the position of the shoreline was constantly changing. Herodotus was also of the opinion that the existing Nile delta had once been under the sea and that the area had been filled in by sediments from the Nile.

Other Greeks taught that the universe was ordered, that it was governed by principles or natural laws, and that continual changes of the earth's surface were the result of natural processes. Among these perspicacious students of nature was Heraclitus, who noted that nothing on the face of the earth was lasting because changes were continually going on through the agencies of nature's processes. Both Pythagoras (born about 582 B.C.) and Empedocles of Argigentum (492–432 B.C.) passed on much the same views in their teachings.

First Principles

The culmination of Greek thought concerning natural phenomena is probably to be found in the lectures and writings of Aristotle, who lived from 384 to 322 B.C. Aristotle, learned and extraordinary pupil of the philosopher Plato, founded his own center of learning, the Lyceum. There, he disseminated his knowledge in polished, thought-provoking lectures. He, as did naturalists before his time, induced from his observations of nature's processes and belief in their uniformity that the existing positions of land and sea were not permanent but that certain land areas had been covered by sea at one time.

It is interesting to note that both Plato and Aristotle, although leading scholars of their time, were not cloistered in their temples of learning; they both had some commercial interests in addition to scholarly ones. Plato sold oil in Egypt, and Aristotle managed what may be considered a drugstore.

From early folklore and the recorded observations of Greek naturalists sprang the fibers that were woven into the principle of the uniformity of nature's processes. Rome, with its Republic, armies, and scholars, followed Greece, with its culture and learning. Leaders of Roman scientific inquiry inherited Grecian theories—by then already in a somewhat decayed condition, to be sure—and were influenced by them. Among many Greek-influenced Roman naturalists were Lucretius (98–55 B.C.), Strabo (active about 7 A.D.), and Pliny the Elder (23–79 A.D.). They came to the same conclusions as had Aristotle and other Greek scholars concerning the past positions of land and sea, and changes in relationships between them, as well as the effects of natural processes in general and their relative long-term uniformity. Strabo indicated that explanations for natural phenomena should be sought among events of daily occurrence or those of obvious character, such as volcanic eruptions and earthquakes. He pointed out that following a volcanic eruption at sea, an island had been formed where before only water could be seen. Pliny

pursuer of nature's secrets, died in his pursuit
...esuvius erupt and its issue of lava and cinders
...ompeii. He left to following generations of natu-
... comprehensive compendium of his observations of
...re entitled *Historia Naturalis* which, despite being a somewhat rambling and disorganized presentation, was the most inclusive work available on natural history for more than three centuries.

During the long span of the Dark and Middle Ages, which followed the decline of the Roman Empire, scholarly activity was limited and its records are primarily to be sought in the monasteries and cloister schools of the time and in the writings of the Byzantine and Arab scholars who borrowed ideas from classical Greek and Roman writings, but only retained them without advancing from them. Advances that were made in knowledge remained submerged until the Renaissance.

Leonardo da Vinci (1452–1519), a versatile genius, was one of the men who during the Renaissance revived interest in observing natural processes and relationships and using these observations to interpret the meaning of phenomena related to past events. He stated, for example, that sea shells visible in the rocks in the Apennines in northern Italy had belonged to animals living in a sea that had once covered that area. River muds from Alpine lands had been transported into that sea and had filled the shells of marine animals; the shells were preserved as the muds hardened to rock. He noted that rivers erode their valleys and deposit fine muds at their mouths, and that animals and plants tend to be buried in the muds. He reasoned that changes occur in shells of animals—that they become petrified—as the muds harden to rock. Areas of rock then rise, or are shoved, above the sea level to become land.

During da Vinci's time and shortly after it, two opposing views concerning earth history confronted one another. One was from the Book of Genesis: the portrayal of the complete formation of the earth a few thousand years earlier during the Creation. Only one catastrophe had wrought any change on the earth, and that had been the Noachian Deluge. After that, the earth had been in a stable state. In this view any change in sea position had to be related to the Deluge.

In opposition was the stand taken in the writings of the Greeks and Romans that changes were always taking place on the earth's surface. Da Vinci renewed this idea with his observations concerning the deposition of muds by rivers and the conclusion that the muds harden to rock and become land.

Along with this difference of opinion concerning change versus stability of the surface of the earth, another controversy raged: a controversy over the nature of the remains—that closely resembled shells of animals then living along the existing shoreline—found in rocks. Violent polemics ensued in the course of debate. That the remains were similar to shells of living marine animals was quite apparent, but how did shells get into rocks? This point was of great concern. To many it seemed that they must have been formed there during the Creation, or had been transported there during the Deluge. There could be no other answers for most people because the teachings from the Book of Genesis held them tightly. For those who were tempted to express some doubt, there was the example of Giordano Bruno who was burned at the stake in Rome in 1600 for stating that there had never been a Deluge and that the positions of land and sea had changed many times. Bernard Palissy was another who had encountered the wrath of the clergy. He was denounced as a heretic for his statements that fossils were the remains of once-living animals and plants. Only gradually and over the time span of two centuries did the aspect of strong resemblances to living organisms outweigh adherence to medieval tenets and were fossils defined as the remains of once-living organisms.

In the years following da Vinci's observations and conclusions concerning nature's activities, a number of observant naturalists began to study natural processes and to discuss their effects on the earth's crust. Among them were Thomas Burnet and Robert Hooke. Burnet, writing in the 1860's, noted that in time all the mountains of the earth could be washed into the sea by the everyday processes of rainfall and runoff and wind abrasion. Hooke pointed out that such commonplace events as a river's washing sediment to the sea, the sea's pounding upon a shore, and the wind's abrading the land all wrought changes in the earth's surface. He went on to say that the most powerful forces operative in changing the surface of the earth were earthquakes and volcanic eruptions. The action of volcanoes had made areas that were once plains into mountains; earthquakes could have caused the foundering and disappearance of portions of land. He explained that if any areas had disappeared in this way, the organisms living on them would also have been lost. Some animals would have been destroyed during such changes. He believed others would have come into existence. Because he considered that extinctions and appearances of new life forms had taken place, Hooke suggested that fossil forms might be used to determine a record of past ages. Hooke was a firm advocate of the theory that the fossils in rocks are the remains of once-living organisms; in holding this view, and certain others, he was among the minority of his time.

Close to a century after Hooke's time, the French naturalist Georges Louis Leclerc, Comte de Buffon, in his *Natural History*, the first volume of which appeared in 1749, attempted to explain the earth as a system of matter in continual motion. He stated (1781, p. 34):

> . . . we must take the earth as it is, examine its different parts with minuteness, and, by induction, judge of the future, from

First Principles

what at present exists. We ought not to be affected by causes which seldom act, and whose action is always sudden and violent. These have no place in the ordinary course of nature. But operations uniformly repeated, motions which succeed one another without interruption, are the causes which alone ought to be the foundation of our reasoning.

Buffon thought that the earth's surface had been formed beneath the sea by tidal and current action. He thus emphasized that commonplace natural events, not catastrophic ones, were the fundamental causes of the surface features of the earth. Buffon, although he was vague as to some of the details, was close to Huttonian thought in regard to uniformitarianism.

Jean Guettard, in the mid-eighteenth century, and Nicolas Desmarest, shortly afterwards, noted that remains of once-active volcanoes were present in the French province of Auvergne. Both stated that at one time volcanoes similar to modern ones had been active in certain areas, such as Auvergne, no longer identified with volcanic activity. Desmarest went further in his studies to establish the fact that volcanic rock in Europe had formed at different times in different ways. He too affirmed that the processes of nature had acted in the past in the same manner as they do in the present.

By the latter part of the eighteenth century, a considerable background of observation and conclusion pertaining to uniformity in natural processes through time had been recorded. Up to this time, however, most of the conclusions were tied to Church dogma in some way, no matter how far the facts had to be stretched. A number of writers, among them Voltaire, who were questioning theological tenets, noted this, pointing out that geology was nothing but the handmaiden of the Book of Genesis, so consistent were geologists in paying obeisance to the history of the earth as given there.

By this time many scientific societies had sprung up in England. The first of these had been founded in London, but those established somewhat later in outlying districts also attracted their coteries of good friends and scholars who enjoyed a pleasant evening of good food, good wine, and good talk together. The Royal Society of Edinburgh was one of these. Late in March of 1785, James Hutton, gentleman farmer and avid student of rocks and minerals, strode into the weekly meeting with more than his usual exuberance: He was to present to the group the results of several years of intensive collecting, sifting, and synthesizing of facts concerning earth processes. His talk, and its printed version, which appeared in 1788 under the title *Theory of the Earth; or an Investigation of the Laws Observable in the Composition, Dissolution, and Restoration of Land upon the Globe* (Trans. Roy Soc. Edinburgh 1:209–304, pls. I–II), encompassed in conclusive form the principle that natural laws may be derived from studying the present processes; that understanding of nature's past operations may be obtained from observations of present natural relationships.

Among Hutton's listeners were James Black, the chemist who discovered carbonic acid; John Clerk, an authority on naval tactics; and John Playfair, professor of mathematics at the University of Edinburgh. These men and the other listeners did not think that Hutton had shaken the foundations of the history of the earth given in the Book of Genesis or that he had questioned Adam's birthdate, which was commonly taken, in keeping with Bishop Ussher's scholarly dictum, to be 4004 B.C. They had heard many of his thoughts before, and some of his listeners had even been on excursions with him to see proofs of his hypothesis. Not until nearly five years after its publication was Hutton's hypothesis attacked. When criticism came, it was sharp and cutting, and pointed primarily to divergences from theological beliefs. The critics then jumped from the particular divergences to branding the entire hypothesis as antitheological, and Hutton as an atheist.

First Principles

To set forth his evidence more fully, Hutton expanded his original talk into a two-volume book, *Theory of the Earth*, which was published in 1795. In the first volume of that work he stated (1795, p. 19):

> In examining things present we have data from which to reason with regard to what has been; and from what has actually been, we have data for concluding with regard to that which is to happen hereafter.

Hutton died only two years after his volumes appeared. His friend and colleague in the Society, John Playfair, undertook the task of amplifying and defending Hutton's views. Hutton was facile with his explanations while talking, but he was not as at ease with the written word. His work was difficult to read and encumbered with excessive verbiage. Playfair was able to summarize and elucidate the Huttonian hypothesis brilliantly; because of his clear, concise style, Hutton's views became widely known. Playfair published his summary of Hutton's work as *Illustrations of the Huttonian Theory of the Earth* in 1802.

Hutton was no idle amateur naturalist, but a man who devoted nearly a lifetime to study of nature. He was educated as a physician but did not pursue medicine as a career. After a brief excursion into chemistry he turned to agriculture, which occupied his activities for a decade and a half, after which he devoted his full attention to study of the earth and, in particular, rocks and minerals. Hutton read widely, and traveled extensively about the British Isles, always observing the rocks and the natural processes acting upon the earth's surface. Never assuming, always letting conclusions come from observable facts, he induced conclusions from his observations. Hutton offered many proofs for his hypothesis that past changes of the earth had been brought about by the agencies of everyday, observable processes. He noted that layered rocks were composed of detrital fragments, grains from preexisting layered rocks, or bodies of even greater antiquity. Running water sculptured the surface of the earth and ultimately deposited the detritus it had carved from the land into the sea. Pressure and subterranean heat caused the formation of stratified rocks from the original deposits of loose detritus on the sea floor. Then, by the operation of great forces, the layers were cast up to form new lands. Once the lands had emerged from the sea, destructive agents—running water, tides, winds—conspired to carry particles of it back to the sea. Hutton (1788, p. 304) maintained: "The result, therefore, of our present enquiry is that we find no vestige of a beginning, no prospect of an end."

Hutton's most breathtaking point was the immensity of time envisaged in his hypothesis. The eighteenth-century mind was scarcely able to comprehend that past time had been so long and that changes had been going on throughout it. To many people who had been taught that the history of the earth spanned only 6,000 years, the thought of a much longer time was inconceivable.

Hutton's conclusions began to be used by a number of naturalists, but none had so broad an influence on the thought of his time as Charles Lyell. Lyell's *Principles of Geology*, published initially in three volumes that appeared in 1830, 1832, and 1833, covered the complete scope of geology, and not only supported but also amplified the principle of the uniformity of natural processes through time by extensive coverage of erosional and depositional processes. Lyell noted (1830, p. 70): "Hutton labored to give fixed principles to geology, as Newton succeeded in doing to astronomy." *Principles of Geology* went through many editions and was widely read by nineteenth-century naturalists. This work did much to exemplify and to furnish compelling evidence in favor of the validity of Hutton's conclusions concerning the long-term uniformity in nature's processes.

All phenomena that are related to the past history of the earth are dependent upon the principle of the uniformity in nature's processes through time for their interpretation. Everything from interpreting shells preserved in rocks as remains of once-living organisms to ascertaining the passage of time by using decay rates of unstable isotopes such as potassium 40 and carbon 14 depends on this principle. For example, the C^{14} method depends on this principle as expressed in the assumptions that cosmic radiation has been of the same intensity for at least the last 35,000 years (the length of time for which this method is most effective), and that the decay rate of C^{14} has always been the same as it is now. Obviously, without a principle of uniformity in natural processes, age determinations based on C^{14} decay could not even be considered.

Hutton's conclusions and Lyell's amplification of them are the heart and soul of geology. As may be noted from this short history of the development of Hutton's conclusions, they have been considered, tested, and used for centuries. Their roots are as old as, if not older than, those of nearly any principle applicable to natural phenomena; they form one of the most broadly based and reliable set of guidelines in scientific investigation.

Superposition

At the foundation of the understanding of the proper sequential order of rocks and the fossils contained in them lies the principle of superposition. That many rocks were layered, and that such rocks commonly bore fossils, were observations made by scores of naturalists from the beginning of man's investigation of the natural phenomena about him. Some rock layers were tilted only a little. They could easily be seen to lie one on top of the other. Other sets of layers were contorted. Their sequential order was more difficult to establish.

Nicolaus Steno studied the rocks of Tuscany in northern Italy, and these studies led him to induce the principle of superposition. In any sequence of flat-lying rock strata, according to this principle, the oldest layers are at the bottom and the youngest at the top.

Steno was his Latinized name; he was born Niels Steensen in Copenhagen. He studied medicine and became physician to the Grand Duke of Tuscany. While in that position, he traveled widely in northern Italy. His inquisitiveness about nature led him into the mountains, especially in the vicinity of Florence, where his studies included the rocks and minerals he found there. He summed up his geologic observations and conclusions in 1669 in *De Solido Intra Solidum Naturaliter Contento Dissertationis Prodromus.* This work was intended to be preliminary to a more extensive study, but that never materialized as theology superseded natural phenomena in Steno's interests: He became the Apostolic Vicar of Northern Germany and Scandinavia.

The *Prodromus* is a classic in geologic literature, and although only a few copies of the original version appear to have been published, it was translated into English and French shortly after its publication in Latin. Steno pointed out that in stratified rocks: (1) a given layer could form only on a solid base; (2) when any layer was forming, either its sides were bounded by a solid substance or else it covered the entire earth; (3) when any layer was forming, the one beneath it must already be solid; and (4) when any layer was forming, only the fluid from which its particles came was above it and therefore no overlying layers could have been present when the lowest layers were formed. The lower layers must be older than the upper in any sequence of strata. Steno also stated that when any series of strata are formed, they must be nearly horizontal. Tilted and deformed strata are the result of displacement by earth movements after deposition. He pointed to two causes of earth displacement,

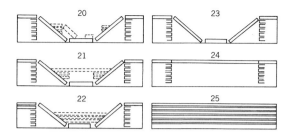

FIGURE 1. Six diagrams used by Steno to illustrate his ideas concerning the sequence of events that resulted in the positions of different groups of strata that he was able to observe, lying at various inclinations, in outcrops in the Arno Valley near Florence, Italy. Steno concluded, using his principle of stratal superposition, that many strata were formed in a horizontal position one above the other by deposition of sedimentary particles from a fluid medium (illustration 25). After deposition, Steno suggested, the strata under the stratigraphically highest were eroded away, leaving a cavity beneath the highest stratum (illustration 24). Then (illustration 23), the highest stratum collapsed into the cavity beneath it. The sea invaded the valley formed by the collapse, and more strata were formed upon deposition of particles from the sea (illustration 22—the new strata are depicted by dashed lines). Steno believed that once again a cavity was eroded from beneath the new stratigraphically highest stratum (illustration 21), and that that stratum fell into the cavity (illustration 20). Diagrams 20, 21, and 25 illustrate Steno's comprehension and discussion of the principle of superposition. [Adapted from N. Steno, *Prodromus*, Florence, 1669.]

volcanic eruption, and caving-in of the surface where substances beneath had withdrawn.

From Steno's almost self-evident principle, geologists could work out local successions of strata with confidence that the lowest were the oldest. Through time, a set of criteria was established by which the tops and bottoms of beds could be ascertained in badly deformed and strongly tilted strata. Geologic structures could then be established and proper superpositional order demonstrated in any area.

CHAPTER THREE **Attempts at Division**

Before turning to the principles that led to erection of a set of time units based upon a succession of fossil aggregates, a succession that resulted from organic evolution through natural selection, let us look at some attempts to divide the rocks of the earth's crust using only the principle of superposition. The best of these attempts led to descriptive generalizations, and the worst to faulty hypotheses.

During the seventeenth and early eighteenth centuries, many rock successions were recognized, principally around areas of mining activity. Among the first published descriptions of such a succession was that of the Englishman John Strachey. In 1719 he wrote about a sequence of strata in the coal-rich district of Somerset in the southwest of England. He noted that the coal-bearing beds were inclined, and that beds of marl and oolite lay almost flat on top of them. In 1725 Strachey published the entire sequence of rock units known at that time in the southwest of England, from the coal-bearing strata through the red marl, lias, and oolite to the chalk at the top. Strachey was able to establish a succession of strata overlying the coal beds that could be recognized throughout a relatively wide area.

Attempts at Division

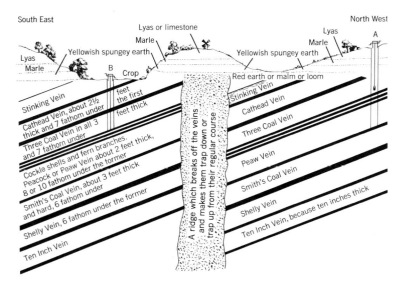

FIGURE 2. Strachey's section through the coal beds and other strata in Somerset, south of Bristol, England. [Adapted from John Strachey, *Philos. Trans. Roy. Soc. (London)*, vol. 30, pl. 2, 1719.]

Then, during the early eighteenth century, geologists began to turn their attention from local and regional stratal successions such as Strachey's to weaving a history of the whole earth from them. As they did so, they realized that older rocks, in general, formed the cores of the higher mountains, and that younger ones formed terrain of lesser elevation. This realization led to the supposition that all of the rocks of the earth's crust might be grouped descriptively into those that form the highest mountains and those that form successively lower mountains. A school of thought developed which stated that as all the oldest rocks were at the cores of the highest mountains, as many of those rocks appeared to be crystalline, and as the rocks constituting the mountains of lesser height were commonly detrital, therefore all of the oldest rocks, the crystallines, formed at one time, and all the detrital rocks at another, later, time. This point of view was developed initially by Johann Lehmann in Germany and by Giovanni Arduino in Italy and was later developed to an extreme by Abraham Werner and his students. Werner's elaboration of the Lehmann and Arduino ideas was widely circulated among geologists and accepted by many at the same time that Hutton, Playfair, and Lyell were demonstrating the validity of the principle of the uniformity of nature's processes through time. Hutton's conclusions stood in opposition to the proposals of the Lehmann-Werner school of thought, whose view of earth history was not based upon a principle induced from observations of actual rock relationships, but rather on an assumption stemming from a very generalized view of the relationships between rocks forming the high mountains, and those of lesser height. This descriptive grouping of mountains was thought to reflect a set of mountain types formed of particular bodies of rock that represented steps in the development of the earth, and, therefore, particular spans of time during earth history. To those who subscribed to this, nothing more was needed to ascertain a period of past time than to recognize a rock type. Any rock body of any magnitude was thought to have formed at the same time in all areas in which it is now found, and any rock body was considered to be of the same age throughout. Obviously, observable processes of nature operative upon and in the earth were ignored in this view, but it was the widely accepted eighteenth-century descriptive approach to dividing earth history into chapters: At the present time it still has its adherents.

During the seventeenth, eighteenth, and early nineteenth centuries, mining geology played an important role in the economy of many countries in western Europe. As a result, much of the geologic work of that time was related in some way to mining, and much of the thought about geologic phenomena was developed by geologists connected with the mining industry. Because it was important, mining schools were established to improve training in mining methods and knowledge of minerals.

Johann Gottlob Lehmann taught mining and mineralogy at one such school in Berlin. He read widely concerning geology and carried on a considerable program of painstaking field work, studying the rocks in Thuringia in great detail. He amassed a wealth of data about the rocks in that area, and learned much about rocks and their relationships in many areas in Germany. His researches led him to propound a descriptive grouping of rocks of the earth's crust. This first published account of a rock-group classification that would be more than just local was incorporated in his unpretentious little book, *Versuch einer Geschichte von Flötz-Gebürgen,* published in 1756. This work set the stage for later efforts to formulate widely applicable descriptive rock divisions.

Lehmann recognized three classes of mountains (and of the rocks that he believed formed the mountains). The first class included the high mountains such as the Alps. Those mountains were, Lehmann believed, composed of strata varying in attitude from inclined to vertical. The rocks were less varied in type than those of the other classes and did not contain organic remains, but they did abound in useful minerals. Because ore minerals were so abundant in these rocks, Lehmann termed this class *Gang-Gebürge.*

Lying on the *Gang-Gebürge* rocks were those of the second class of mountains, which Lehmann called the *Flötz-Gebürge.* Lehmann believed that the rocks that formed the *Flötz-Gebürge* (he called these rocks the *Flötz-Schichten*) had formed during the Noachian Deluge. Waters had fallen on the high mountains during the Deluge, mixing with the earth and sweeping up the animals and plants that had been living on the slopes. As the waters receded, sediments, containing organic remains, were deposited in the horizontal or near horizontal position on the flanks of the high mountains. These sediments hardened into rock layers, entombing the organic remains. Lehmann noted that the organic remains had been preserved in different ways. Some had petrified, others had mineralized, and yet others had rotted away leaving only imprints on the rock matrix. The high mountains had been denuded of their soil cover during the Deluge; it had formed the sediment that had been deposited and hardened into the layered rock (the *Flötz-Schichten*) that constituted the *Flötz-Gebürge*. The high peaks that rose above the rocks of the *Flötz-Gebürge* were thus commonly bare and craggy. Lehmann pointed out that the *Flötz-Schichten* yielded economically useful coal and marble and some minerals.

The third class of mountains had formed since the Deluge, Lehmann believed, by accidents of nature, such as volcanic eruptions, storms, landslides, and earthquakes, that had occurred from time to time. The rocks that were included in this third class of mountains were loosely consolidated.

It may be noted that Lehmann thought of the Gebürge not just as topographic entities—mountains—but also as that sequence of events that had acted collectively to form the mountains. Thus to Lehmann, a Gebürge included a set of events that were essentially unique in time. It is thus easy to comprehend how the Gebürge were thought to be significant great events in the history of the development of the earth and how the rock types that constituted them were also thought to be unique in time.

Perhaps the most lasting contribution in Lehmann's book was his accurate and detailed description of the *Flötz-Gebürge*. The rocks he studied are now known to be Permian in age. He presented a number of measured sections indicating the order in which the strata followed one another, and he included some diagrams in which the relationships were graphically depicted.

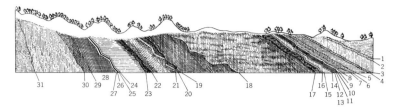

FIGURE 3. Diagram used by Lehmann (1756) to illustrate the succession of stratified rocks he classed as Flötz-Schichten (numbers 1-30) overlying the rocks of the Gang-Gebürge (number 31) in the vicinity of Mansfeld on the eastern margin of the Harz Mountains in Germany. Layers 1 through 17 include various limestones and shales including copper-rich shale. Layers 18 through 23 include red sandstones and shales. Layers 24 through 30 include coal and interbedded shale. This succession, in terms of the present time scale, is Permian in age. [Adapted from J. G. Lehmann, *Versuch einer Geschichte von Flötz-Gebürgen*, Acad. Wiss., Berlin, 1756.]

Many of the rock terms he used were adopted from the miners. Some, such as *Zechstein,* or mine stone, and *rothes Todtliegendes* (commonly written, *Rotliegendes*), or red lower layer, are still in use.

The general history of the earth as discussed by Lehmann was a work of its time. During the eighteenth century, belief in a Deluge as the one great catastrophe inflicted upon the earth during its history and belief in a creation at some time before that were the only acceptable explanations concerning the birth and development of the earth. Lehmann did note that mountains of the third class had formed during accidents to the earth's surface, indicating some appreciation of the effects of naturally occurring phenomena that had taken place since the Deluge. His reckoning of the earth's history followed from his careful study in an area familiar to him. His proposals were soon followed by those of others who used his, to some extent, as models.

A contemporary of Lehmann less often cited and less well

known but equally astute in his observations and conclusions was George Christian Füchsel. Dr. Füchsel had a life-long enthusiasm for rock and mineral collecting. He was physician to the Prince of Rudolstadt in Germany, where he spent most of his professional life. His geologic avocation directed his attention to the ancient rocks in Thuringia, a short distance south of Rudolstadt. Rivers had cut through the low mountains there, revealing layered strata now known to be of Permian and Triassic ages. They were seen to overlie older rocks. Füchsel published his observations in two works. The first, entitled in Latin *Historia terrae et maris ex historia Thuringiae permontium descriptionem erecta* (*A History of the Earth and the Sea Based on a History of the Mountains of Thuringia*) appeared in 1762 and included a geologic map of the territory in which Füchsel had studied. The second book, published in 1773 and entitled *Entwurf zur ältesten Erd und Menschen Geschichte*, further amplified his opinions.

Füchsel had studied the rocks in Thuringia for twenty years or more before presenting his views of the history of the earth and the details of a sequence of stratified rocks. He considered that the layered rocks had been formed on the sea floor, and that the highly inclined rocks that appeared to rise from beneath them, making up the high mountains, had been formed from an even older set of marine sediments. Certain parts of these rocks were considered to have fallen unevenly into the interior of the earth, thus causing them to be inclined from an originally horizontal position.

Füchsel's most lasting contribution to the geology of Thuringia was his careful description of nine rock units. From the highest or youngest to the oldest in the succession, these are:

9. Muschelkalk—the present day Middle Triassic
8. The sandstone series—the present day Lower Triassic, Bunter Sandstone

7. Granular limestone and marls—now a part of the Upper Permian Zechstein
6. The metalliferous rocks and copper slate—now a part of the Upper Permian Zechstein
5. White rocks with sand and clay
4. Red rocks
3. Slate with marble lenses
2. Coal-bearing rocks
1. Basal rocks that were inclined and formed the peaks of the highest mountains in Thuringia

In addition to describing the rocks in detail and presenting sections and the geologic map, Füchsel also described fossils that are characteristic of some of the rock units.

Füchsel traveled little, and his writings were scarcely read during his time. Only after his death did his work receive much attention. Füchsel believed that all the rocks he saw had been horizontal at the time of deposition and that, during any one time of deposition, conditions were so similar that the rocks deposited then were homogenous in composition. Rock bodies so formed indicated to him major events or epochs in the history of the earth.

At about the same time that Lehmann and Füchsel were working out successions of layered strata in Germany, Giovanni Arduino was studying the rocks and mineral deposits in Tuscany. This was, at least in part, the same territory that Steno had examined approximately a century earlier while formulating the principle of superposition.

Arduino was an inspector of mines in Tuscany at the start of his career, and later became Professor of Mineralogy at Padua. He was an energetic writer and had a great deal of influence among fellow geologists and mineralogists in Italy and also on visiting geologists who came to Italy to study. Because of this influence, Arduino's geologic conclusions received wide reputation. The majority of his stratigraphic ideas that are pertinent to our discussion were published in the 1760's and 1770's.

Arduino (as had Lehmann a few years before) recognized three kinds of mountains, that is, mountains formed of three classes of rocks, which he named the *Primitive,* the *Secondary,* and the *Tertiary.* The Primitive rocks included the unfossiliferous schists, granites, and basalts that formed the cores of high mountains. The Secondary rocks were richly fossiliferous marine limestones, marls, and clays, and some other sedimentary rocks as well. The Secondary rocks were layered, and were found on the flanks of the high mountains. The Tertiary rocks consisted of a younger group of limestones, sandstones, marls, and clays, all abundantly fossiliferous. Arduino noted that, in some cases, materials constituting certain of the Tertiary rocks were derived from those of the Secondary. Tertiary rocks were seen to form low hills and mountains. In addition to the three main classes, Arduino also noted a fourth, the volcanic group. The rocks of this group included lavas and tuffs interbedded with fossiliferous marine strata. He concluded that this volcanic group of rocks had accumulated as the products of volcanic eruptions alternating with inundations by the sea.

Arduino believed that the sequence—Primitive, Secondary, Tertiary—was correct, but he maintained that this sequence did not imply that all Primitive deposits had formed at the Creation and all Secondary deposits had formed during the Deluge. To Arduino, the divisions were generalities. Geologists had to determine the order of local rock units and their relative relationships. To determine that a rock was a limestone was not enough to establish its position in the general rock succession. It might equally well belong in the Secondary or in the Tertiary.

The entire domain of geologic thought during the latter half of the eighteenth century was dominated by the influence and authority of one man, the renowned Abraham Gottlob Werner,

Attempts at Division

Professor of Mineralogy at Freiburg, Germany. Werner published little but was a powerful and enthusiastic lecturer who had the ability to impress his opinions indelibly upon the minds of his listeners. He had a wide range of knowledge of geologic subjects. He was kind and helpful to his students, and built up a devoted and loyal following who eagerly accepted his every word: He became a sort of geologic oracle, and his lectures, the latest dictates. His opinions were regarded by many as the final decisions—whatever the subject. Werner appears in historical retrospect as a curious blend of a kind, energetic teacher capable of stirring within students a zeal similar to his own for geology, and at the same time, a dogmatic theorist blindly intolerant of any view but his own. His pupils were taught only one view of geology—his. Later in life he refused to debate his ideas, or even to read the current journals to bring himself up to date in his field.

Werner was born in 1749, the descendant of ancestors who had for centuries been engaged in the iron industry in Germany. His father was interested in minerals, and young Werner from his early childhood had minerals as playthings. He worked in the foundry with his father for five years before the desire to learn more about minerals led him to the then recently founded Mining Academy at Freiberg. He studied mining there. He then went to the University of Leipzig where he spent three years studying law and mineralogy, and then he accepted an invitation to join the teaching faculty at the Freiberg Mining Academy in 1775. He spent the rest of his life there, and changed the reputation of the small academy from that of a training school for local miners to that of a center of learning known to geologists throughout Europe.

Werner's fame spread primarily through his diligence as a teacher and the scope of his lectures. There was nothing he would not do for his pupils. Often, if the room became too crowded, he would divide the group into two parts, and give the lecture twice. His mineral collection and his library of works on minerals were shared willingly with his students. His stirring lectures claimed the whole of the student's attention; they covered not only mineralogy but also the entirety of geology and its effect on man. Unhappily, Werner was disposed to teach dogmatic theory and speculation with little regard for facts and apparently little, if any, regard for demonstrable principles. His ideas on the history of the earth were based primarily upon assumptions that in his mind became facts and were passed on as such. Seldom in Wernerian training were observations gathered and conclusions drawn from them. Indeed, in his later years, Werner did not venture far to study rock relationships in outcrop. Surprisingly, this great teacher turned his back on studying relationships at the outcrop, preferring to proceed on blind faith in his beliefs. This procedure led him and his followers into heated, often vituperative, arguments with those who had made observations and, reasoning from them, had examined assumptions.

Werner's initial sketch of his ideas concerning the succession of rocks that compose the earth's crust was published in 1787 under the title *Kurze Klassification und Beschreibung der verschiedenen Gebirgsarten.* The work was later modified by Werner in lectures, but never in print. Many of the modifications and amplifications of Werner's ideas are in the publications of his students. Werner had studied only the rocks in his native area, Saxony, Germany, when he published his ideas. His knowledge of the rocks of the earth's crust was thus limited by his lack of observations with any significance for generalizing. Nevertheless, he discussed the geologic structure of the entire globe. He doubtless had read the publications written by Lehmann and Füchsel for his teachings were dependent upon certain of

Attempts at Division

their ideas. Werner concluded that the whole earth had at one time been enveloped in an ocean, and that from this ocean all the first-formed (Primitive) rocks of the existing lands had been deposited by chemical precipitation and that nearly all rocks were marine deposits. Because the ocean surrounded the globe, the earliest formations similarly surrounded it when they were deposited, and they succeeded one another in a definite orderly arrangement. Water was, to the Wernerian school of thought, the agent of foremost import in the history of the earth.

Werner followed Lehmann in stating that the Primitive rocks were the first formed, the first chemical precipitates from the earth-embracing ocean. They included granite, which Werner maintained was the oldest rock, gneiss, slate, basalt, and other old-looking rocks. Above the Primitive rocks, Werner recognized the *Transition* rocks, which contained some organic remains, suggesting that they had been formed about the time that life first appeared. These rocks were primarily chemical precipitates, but did include some materials deposited by running water, which indicated that the waters of the universal ocean had started to lower and the Primitive rocks had been eroded. To Werner, the *Flötz-Schichten,* or stratified rocks, followed above the Transition. The Flötz rocks were primarily deposits from running water, but they contained some chemical precipitates. Included in them were sandstone, coal, limestone, and slate. Fossils were noted to be common in the Flötz beds. After prolonged retreat of the sea, the fourth type recognized by Werner, the *Alluvial* rocks or *Aufgeschwemmte-Gebirge,* were formed by the action of running water coursing over the land, sweeping up loose materials, and depositing them in the sea, and by volcanic activity in which dust and cinders were spewn out. The Alluvial rocks included clay, peat, ash, and cinder beds.

To Werner, the succession of the four main types of rocks

revealed the four primary steps in the formation of the earth's crust. These steps were simply facts to Werner, and not working hypotheses.

An objection to this neat package of earth history was posed by asking what happened to the water as it receded. Werner never answered the question, and neither did any of his disciples. Another objection to Werner's conception of the history of the earth was that many outcrops were known where the strata were out of the sequence proposed by him. This difficulty was easily dispensed with by calling up the ocean once again to inundate a part of the globe and precipitate certain rocks, and then to roll away to distant and unknown regions. In some places, the rocks that were out of the proposed sequence were at a much higher elevation than they were supposed to be. The ocean had to be brought back to a considerable depth in these cases to sweep over mountainous areas from which it had receded. Deposits then had to drop from it. Such manipulation of the sea could not be tolerated by most scientific thinkers of the time. Werner, in addition to ignoring the problem of how the sea might have invaded and withdrawn, also ignored an abundant literature presenting the rudiments of chemistry as well as the works of Hooke, Steno, and a host of others in his own field that discussed such earth movements as earthquakes and volcanoes, which forced strata up and rent them apart.

Werner ignored the methods for establishing facts as well as the principles formulated and tested by his predecessors. He proposed his own ideas based primarily upon assumptions. So eloquent was his spoken word and so kindly a man was he, that generations of students of earth history blinded themselves to the possibility of scientific investigations of the history of the earth.

Despite the shortcomings of Werner's dogma, he did give a large number of students a comprehension that the natural proc-

Attempts at Division

esses of rock weathering and of rock lithification following deposition were natural phenomena that occurred regularly. Werner also contributed to a better understanding of the rock succession of the earth's crust by identifying the Transition rocks and their fossil content. The Transition rocks were later recognized in many places in Europe besides Saxony and provided fertile ground for research for stratigraphic geologists.

Werner's four-fold division of the rocks of the earth's crust became widely known and was strictly employed by many of his students. Others set out to use it but quickly ran into demonstrations of its fallacies. These four major units, or at least some of them, were used in many parts of Europe to classify rocks according to superficial similarity. Werner believed that the rocks themselves could be used to determine time relationships in the prehistoric past. Because each rock type was deposited at a discrete time, identification of any rock type could be used to indicate the particular part of the earth's crust from which it came.

Werner's doctrine ran head-on against the facts observed by other geologists. One of the tenets of Werner's dogma that was particularly offensive to many was that explaining the origin of basalt and other rocks now known to be volcanic. The Wernerian view was that they had been precipitated from the sea at a particular time in the earliest chapter of earth history. Volcanoes, however, had long attracted the attention of a large group of able investigators. Nicolas Desmarest had demonstrated that basalts in France were probably related to volcanic activity. Arduino had noted that basalts were igneous in origin. Then Johann Voigt proclaimed that some of the very samples of basalt studied by the learned professor at Freiberg were not precipitates. This provoked the impassioned battle between Werner's followers, the "neptunists," and those who held that basalt had an igneous origin, the "vulcanists." Werner admitted that volcanic eruptions did occur, but claimed that they had not participated in the construction of the earth's crust. Desmarest and others only said "come and see," to the neptunists, but few came. The noted geologist Leopold von Buch was one who did venture to see for himself and, after studying the cone and crater at Puy de Pariou in the Auvergne area in France, slowly became convinced that basalt did have an igneous origin.

By the late eighteenth century, a basic four-fold division of rocks of the earth's crust was generally recognized. The Lehmann-Werner interpretation favored an absolute relationship between rock type and time in the prehistoric past, whereas Arduino's divisions were more nearly relative for he had noted that several rock types might be found in more than one of his major divisions. At that time, geologists had only Steno's principle of superposition to help them establish rock relationships. Even though the principle of the uniformity of nature's processes through time was more or less recognized, it was not published in definitive form until a year after Werner's publication of his ideas of the history of the earth's crust. Not until 1795, during the height of Werner's popularity, was the principle, as expressed in Hutton's conclusive presentation, circulated widely. Even Hutton's studies included only the dynamics of the inorganic world, and did not include life processes.

CHAPTER FOUR **Darwin's Contribution—Organic Evolution Through Natural Selection**

One event that the reporters of the London *Times* didn't bother to cover on July 1, 1858 was the monthly meeting of the Linnaean Society. What was said there was received very calmly by those who attended, but has come to be recognized as one of the major developments in the history of human knowledge: At that meeting Charles Darwin and Alfred Wallace read their joint paper entitled *On the tendency of species to form varieties, and on the perpetuation of varieties and species by means of natural selection.* Perhaps the *Times* may be forgiven its ignoring a naturalists' meeting for it was a warm, sunny summer day and there was a cricket match on at the Oval. The Darwin-Wallace principle of organic evolution through natural selection has not escaped the notice of biologists and geologists, although the latter have been somewhat backward in grasping its usefulness. Paleontology has, of course, not only supplied proofs that evolution has occurred, but has also made significant contributions to understanding the processes of evolution. Stratigraphic geologists have, however, been slow to realize that evolution through natural selection provides them with a succession of events that are unique in time with which they may determine the relationship of one event in geologic history with others. The unique events in time also provide markers or points in the general succession of fossils that may be obtained from the rocks and used as base-marks for time units.

Because the principle of organic evolution through natural selection is so important to the reckoning of past time by geologists and because its importance has not been widely understood, narration of the historical development of the geologic time scale will be broken at this point to include a discussion of organic evolution.

The succession of faunas and floras seen in the rocks of the earth's crust is the product of several factors of which the most important is the evolution of organisms through natural selection. Evolution thus is the very basis of the geologic time scale although the scale itself was erected before Darwin and Wallace presented their principle of natural selection to the scientific world.

The principle of organic evolution through natural selection has its roots in man's observation of life about him. The first foreshadowing of the principle of which there is a record is in the ideas of Empedocles, who lived in the fifth century before Christ, and taught that nature produced many organisms, with the unfit or less fit being eliminated. After the early explanations of the Greeks and Greek-influenced Romans, little in the way of development of this principle can be noted until the Renaissance.

During that period, many observant students of nature pointed the way toward comprehending fossils as the remains of once-living animals and plants and noting that these remains changed in form from bottom to top in any succession of layered rock. Buffon, in the latter part of the eighteenth century, revived Classic Greek thought on organic evolution, and suggested, in a germinal form, the natural selection principle that Darwin and Wallace successfully demonstrated a century later.

Buffon maintained that heavy, delicate, or poorly defended animals had disappeared, or would, from the earth. The weak forms would be eliminated by the strong. He also asserted that animals became adapted to a particular geographic area with a certain climate and food, and that when they left their natural habitat, their appearance slowly changed. The changes were passed on to their descendants.

Shortly after Buffon's most active days, Baron George Cuvier and Chevalier de Lamarck rose to important positions in the scientific world of the latter part of the eighteenth century. Because of his high professional reputation, his gracious manners, and his stirring lectures, Cuvier became the more popular and the more widely acclaimed. He was showered with honors, among them the titles "Baron" and "Peer of France." He was Professor of Anatomy at the Paris Museum of Natural History, and received considerable renown as a specialist in vertebrate anatomy. Cuvier concluded from his studies that all organisms now living had been created at one time, and that they had not changed significantly since their creation. He did admit that certain variations from the original forms had occurred, but that these were only slight modifications within fixed limits (the limits being essentially fixed by the Creator).

In his professional capacity, Cuvier felt called upon to discourse on the history of the earth and life upon it. This he did in *Essay on the Theory of the Earth*, which was translated into English from the French by the Scottish geologist Jameson. Because of Cuvier's popularity and reputation, as well as his strict adherence to Biblical tenets, his catastrophism ideas of earth history and of organismal changes found ready acceptance among naturalists of his day. Cuvier thought that several revolutions had been produced on the surface of the earth by great catastrophes and that these had led to profound changes in the nature of the rocks and of such fossils as they contained.

He noted that the presence of certain fossils in some rocks permitted the conclusion that such rocks had been deposited from a fluid. Cuvier thought that fossils were the remains of once-living organisms that had been annihilated during the catastrophes. New organisms were created following each catastrophe so that the succession of catastrophes killed off one set of creatures and a new set was created after each catastrophe. The catastrophes were concluded to have been sudden in occurrence. The last of them had been the Noachian Deluge. Before that one, several other catastrophes had occurred, each killing off life that existed before it with the exception of those that occurred before life had been created. Because life was created anew after each catastrophe, each new creation was characterized by new associations of organisms.

Cuvier chided geologists for not collecting fossils in superpositional order because he noted that fossils were essential to establishing a history of the earth. In this regard, he said (1817, p. 54):

> The importance of investigating the relations of extraneous fossils with the strata in which they are contained, is quite obvious. It is to them alone that we owe the commencement even of a Theory of the earth; as, but for them, we could never have suspected that there had existed any successive epochs in the formation of our earth, and a series of different and consecutive operations in reducing it to its present state. By them alone we are enabled to ascertain, with utmost certainty, that our earth has not always been covered over by the same external crust, because we are thoroughly assured that the organized bodies to which these fossil remains belong, must have lived upon the surface, before they came to be buried, as they now are, at a great depth.

Chevalier de Lamarck, first Professor of invertebrate zoology at the Paris Museum of Natural History reached conclusions different from Cuvier's in regard to organisms. He asserted that organisms responded to changes in their environment by adaptation.

Environments were constantly changing, hence organisms were continually trying to adapt to changes. Through such adaptation and over long periods of time, Lamarck maintained, complex organisms developed from simple ones. The effect of environmental change on the anatomy of organisms was Lamarck's keenest contribution. His understanding of adaptation was that an environmental change exerted enough pressure on an organism to make it modify directly: if an environmental change created the need for a new organ, the organism's body fluids would stir and gradually produce a new organ; if an organ already existed and an environmental change proved it useful, it would develop; an organ that was not used would wither away. Lamarck mistakenly assumed that such modifications in organisms would be passed on to their offspring. To Lamarck, use and disuse of organs and the organismal response to needs formed as the environment changed, coupled with the inheritance of the changes, produced, in time, the great variety of organisms inhabiting the earth. His assumptions about adaptation of existing organs in response to environmental changes have been shown to be, in essence, correct; those about inheritance to be mistaken.

Evolution in the late years of the eighteenth century and the early years of the nineteenth was thus considered from at least three points of view: That inherited from the ancient Greeks and rekindled by Buffon in which there was a notion of the continuity of life, that of Cuvier and the special creationists, and that of Lamarck. These points of view were widely discussed in the half century or so before Charles Darwin and Alfred Russel Wallace pointed to natural selection as the basic means by which evolution proceeded. That half century was a time when, as we shall see in the next chapter, there was a veritable maelstrom of discourse on natural phenomena.

By the close of the eighteenth century, all of the ingredients of Darwinian evolutionary thought were at hand. The concepts of natural selection, organismal variation, and environmental change—supported by a fossil record of extinct organisms—awaited synthesis. The synthesis was not made until the mid-nineteenth century when two men working independently of each other, Charles Darwin and Alfred Russel Wallace, induced the principle of organic evolution through natural selection. Darwin had spent nearly thirty years gathering data concerning natural phenomena. He had read widely, including Lyell's *Principles of Geology*, and been impressed by the principle of uniformity in nature's processes. He had also been impressed by the fact that nature enforced a struggle for existence, which he encountered in T. R. Malthus' discussion of populations. Wallace was an able, widely traveled, and well-read naturalist.

The story of Darwin's receiving, in the early summer of 1858, a manuscript from Wallace containing ideas on evolution like his own, and then graciously permitting a joint formal presentation of the evolutionary idea to be arranged is well known. The July 1858 meeting of the Linnaean Society of London was a historic one, for there Darwin and Wallace put the principle of organic evolution through natural selection before the world. Darwin then went on to "abstract" his voluminous notes and thoughts on evolution in his *On the Origin of Species by Means of Natural Selection, or the Preservation of Favoured Races in the Struggle for Life*, which appeared in November 1859.

The Darwin-Wallace principle of evolution through natural selection indicates that all organisms have evolved over a long period of time through gradual change from common ancestors. Darwin pointed out that commercial animal breeders acted as selectors by choosing only the "best" (most commercially desirable) animals to breed. For example, cows yielding the greatest quantities of milk would be selected for breeding, and after successive breedings over a period of time, cows producing even larger volumes of milk would be the end result of the breeder's selection. Darwin pointed out that the effects of such selections would be cumulative because the breeder would constantly choose only the best producers for each breeding. The changes would result from the variability of the individuals within the species being bred and from the selection made by the breeder.

After analysing domestic animals, Darwin turned to nondomestic ones to consider whether a natural force took the part of the selector-breeder. He noted that differences among individuals of the same species existed in nondomestic animals as well as in domestic ones. Some of these variations proved to increase the individual's chances for surviving. Individuals with such "better" characteristics would be the most likely to breed and thus pass on their characteristics to descendants. Darwin also noted that in many species large numbers of individuals are produced, and that they are in competition with each other and with their elders for survival. The incompetent, the unwary, and those to whom accidents occur tend to be eliminated in this competition. What is the net result of this competition? The conclusion Darwin reached was that those individuals most fit for their mode of life and environment would be the survivors. They would be the ones to breed, and to pass on their traits to their descendants. The selection through competition among individuals is a natural one, and Darwin and Wallace termed it *natural selection*. Natural selection takes the role in nature that the selector-breeder does with domestic animals.

Darwin's great contribution to the understanding of organic evolution was the recognition of natural selection as the process by which new species could arise. He amassed a vast wealth of carefully evaluated evidence to support his recognition.

Darwin pointed out that the greater the variation in the environment, the greater the opportunity for organismal variation and thus the greater the possibilities that new species might arise. The more varied environment only grants an opportunity of a wider range in mode of life for individual variants; it does not promote the variants. The individual receives as its heritage the potentialities for variation.

Darwin was unable to state how traits were passed on from one generation to the next. Since his day, the transmission of heritable characteristics and the steps by which evolution proceeds have been, to some extent at least, worked out. Study of genetics has greatly increased comprehension of the mechanics of evolutionary change and the inheritance of traits. Founded on Mendel's experiments with common garden peas, the results of which were published seven years after the first edition of Darwin's *Origin of the Species* but were unnoticed for almost half a century, genetics and biochemistry have pointed to genes, that is, chemical loci on tiny thread-like chromosomes, as the controllers of traits. (Perhaps more precisely, genes may be considered segments of the DNA molecule.) In the study of evolutionary change, different individuals of the same species have been shown to have different combinations of genes. The different gene combinations found among individuals of the same species and the corresponding variable frequency of occurrence of many intraspecific genes may give rise to considerable genetic variation within a single species. Gene frequency within a species is important because the genes with highest frequency will, obviously, be those most commonly passed on to succeeding generations. Evolutionary change proceeds by environmental interaction with the gene compositions of members of species in time and space. A part of a group of organisms constituting a species may become isolated from the others and no longer interbreed with them. The changes accumulated in the isolated population throughout many generations of reproductive history may eventually cause it to become a new species. New species are evolving today, and have been evolving ever since organisms appeared on the earth's surface. Such speciations are particular events that a geologist may use for reckoning time in the prehistoric past.

Darwin noted that although the fossil record was incomplete it revealed many examples from the past of organismal variation and modification. Fossils are, he asserted, the evidence that organisms have changed throughout the past, and that some species have become extinct while others have survived.

CHAPTER FIVE **A Maelstrom of Opinions**

Faunal (and Floral) Succession

With the understanding that it is organic evolution through natural selection that is inherent in the succession of fossils seen in the rocks of the earth's crust, we may now resume our story of the growth and development of a time scale based upon organic evolution. The next step in that growth, after those of the principles of uniformity in nature's processes and superposition, was the recognition of the principle of faunal (and floral) succession. That principle opened the way to interpreting certain fossil aggregates as being unique in time and therefore useful as the bases for time units. The time units founded upon the fossil aggregates in the years after the recognition of the principle of faunal succession were vaguely defined but they were fit together to form a sort of time scale that is widely used today. The units that were established on fossil aggregates represented relatively long durations as they were founded on large groups of fossils, which incorporated many changes in evolutionary development in several different groups of organisms. Most of the units based upon fossil aggregates were recognized before Darwin and

Wallace presented the principle of organic evolution through natural selection to the scientific world. After they did so, it was slowly realized that evolution as Darwin and Wallace outlined it was the reason that there was a succession of faunas and floras.

Before Darwin's time, many collectors noted that the fossils they obtained from stratigraphically lower layers in any succession of rocks were different from those collected from the upper layers. No general principle was induced from these empirical observations until the work of William Smith, British canal builder and geologist. Smith demonstrated the validity of the principle of faunal (and, really, floral also) succession, which indicates that fossils have a definite order of succession that may be determined and that the same fossil aggregates occur in the same sequence wherever they are found.

One of the first recorded suggestions that fossils might be used to indicate ages of rocks was made by Robert Hooke at the end of the seventeenth century. His suggestion lay fallow for the greater part of a century before it was used by the French Abbé, Jean Louis Giraud-Soulavie in 1779.

The Abbé gave most of his attention to the old volcanoes in France, and devoted the greatest part of his seven-volume treatise, *Histoire naturelle de la France meridionale,* to their description. In that work, he described limestone formations of three ages among the calcareous rocks of the Vivarais district. Each of the ages was characterized by a distinct fossil aggregate. Limestone of the first age contained shells of ammonites, belemnites, and other forms that do not have close analogues in animals now living. This limestone underlay all other strata in the area, and was thus the oldest. Above it, the Abbé recognized limestones of two younger ages. That of the second age contained some of the same fossils as the first associated with some fossils similar to existing animals. The third age was denoted by shells similar to those of animals living today.

The Abbé also recognized two more ages of rocks in the Vivarais district. The fourth was slate and shale containing plant remains, and the fifth was alluvial deposits containing remains of trees and land-dwelling animals. Most of the fossils of both these ages were forms similar to existing animals and plants.

The Abbé considered the ages to be actual periods of time in earth history and that because of their superpositional relationships, their relative positions were secure. He also believed that if the sequence that he had recognized in the Vivarais district could be seen in other places, a chronology of the earth based on fossils could be established.

A few years after the Abbé was studying these matters, Cuvier expounded his catastrophism ideas of earth history. He did see a succession of faunas but related them to his catastrophes and the subsequent creations.

There is at least one twentieth-century elaboration upon the Cuvier theme of catastrophism. It views earth history as a series of natural rhythms in mountain making, in advance and retreat of the sea, in sedimentation, and in development and destruction of organisms. Each episode in earth history closed with a worldwide sinking of the ocean basins and a rising of the continents that forced the sea to retreat to the deepened oceanic areas. Continental rise and ocean-basin sinking were periodic events of short duration but pronounced effect. Following each continental uplift came a long period of erosion. The sea advanced as the continents wore down. The spread of vast epicontinental seas across leveled lands took place more or less simultaneously the world around. Similarly, the earth movements that forced the sea to retreat also took place simultaneously all over the world. Because earth movements are considered the key to the entire cycle of events, and because they are considered to have been periodic and worldwide, those who support this explanation assert that earth movements should be the ultimate basis for dividing past time. In this scheme, sea retreat would have a marked effect upon conditions for life, with many organisms dying out at the close of each episode; new organisms would develop as the sea advanced again. So close is this rhythmic view of earth history to Cuvier's "catastrophism" that it might be dubbed "neocatastrophism," although it is hardly as unrealistically neat as was the original.

At the same time that Cuvier's catastrophism was achieving acclaim, a little-known, poorly educated British surveyor was developing an idea that grew out of his engineering experience. William Smith was born in 1769, the same year as Cuvier. He received only limited schooling, and was employed by a surveyor while in his late teens. He became so proficient in his surveying duties that he was soon entrusted with numerous engineering tasks as well. In 1793 he did the survey work for a canal in Somerset, and supervised its construction. This job gave him ample opportunity to work out the stratigraphic succession of the rock layers above the coal beds in that area, and, as fossils were plentiful in many layers, to collect them and observe the aggregates in each layer. The canal was cut through layers that are now included in the Triassic and Jurassic. Study revealed that each rock unit was typified by a definite fossil aggregate.

Smith was elected to membership in the agricultural society at Bath in 1796. He became acquainted with others there who were interested in minerals and fossils. Smith discussed with them his ideas on recognizing rock units by their fossil content. Assured by them that his ideas were unique, he began to set them down on paper. Hindered somewhat by a lack of time and by his poor education, he did not push toward publishing his writing. He had made notes of local stratigraphic sections, he knew how to identify strata by their fossils, and he could determine a particular stratum's position in a succession of strata by its fossils

FIGURE 4. A lock of the Somerset Coal Canal near Bath, England. The width and depth are each about 6 or 7 feet. The facing is in oolitic limestone. The construction of the canal was superintended by William Smith. [Sketched from a photograph supplied by A. O. Woodford.]

even where the relationships were otherwise obscured. Writing was difficult for him, and he realized that he could demonstrate his knowledge in another form. He knew that maps of soil distribution in various areas were published in agricultural reports. If soil distribution could be shown on a map, why not distribution of rock units? Smith pursued this thought with zest, and made geologic maps of the area about Bath which he showed to his friends in the agricultural society at Bath. Smith depicted four units, those now known as the Coal Measures, the Trias, the Lias, and the Oolite, on his first maps, made during the 1790's.

When his employment with the canal-building firm terminated in 1799, Smith undertook a new project that was to be furthered by his taking a series of civil engineering jobs in many different parts of Britain: making a geologic map of England and Wales. He traveled as much as he could, making notes of the rock successions and the fossils contained in each unit as he went. Before long, his notes had grown so voluminous and his idea so exciting to him, that his professional work became only a means to an end—that of completing his geologic map. Even all of the traveling to get him from job to job was not enough: He took many side trips and additional excursions to gather more information into his store of geologic information. During the first few years of the nineteenth century, Smith made small maps on which only a few formations were portrayed. Progress toward publication of the comprehensive geologic map of England and Wales was slow; Smith had little time to devote to the project as he had to continue to work and to gather data, and it was not easy to find money to finance publication. The Geological Society of London, founded in 1807, was approached, but turned its back on the proposal that it support the unorthodox ideas of a common working man.

One of Smith's friends at Bath who was perspicacious enough to judge the merit of his work was the Reverend Benjamin Richardson, who had a large collection of fossils from northern Somerset and who was well read in science and scientific procedure. Richardson had accompanied Smith on several trips to see for himself the succession of strata, each typified by a faunal aggregate, and was impressed with the veracity of Smith's idea. He turned his fossil collection over to Smith and was astounded at the short time it took Smith to arrange the fossils according to the strata from which they had come. Smith assured him that throughout the entire area the same strata were always found in the same superpositional sequence and were typified by the same distinctive fossil aggregates. Richardson and another friend with similar geologic interests, the Reverend Joseph Townsend, were so impressed with Smith's proofs of the validity of indentifying rock units by the fossils they contained, since there was a distinctive fossil aggregate in each unit, that they coerced Smith into dictating a list of the stratal succession in northern Somerset. Details of the lithologic aspect, the fossils that were contained, the thickness, and the localities at which each might be studied were included for each unit in the list. This list was dictated in June 1799 and was widely circulated soon thereafter. It demonstrates that Smith had not only grasped but also had used the principle of faunal succession in a certain area before the year 1800. The Abbé Giraud-Soulavie had recognized that rock layers in a local stratigraphic section were characterized by distinctive fossil aggregates. Smith went further than that by demonstrating that particular stratal intervals could be recognized over a broad area by their unique fossil aggregates. Smith demonstrated to be generally true that which the learned Abbé had shown to be true in a particular section. Using the knowledge of faunal succession that he had so carefully gleaned in his surveying duties, Smith was able to place any rock formation bearing fossils in its proper superpositional relationship by examination of the fossils and comparison of the faunal aggregate with others whose stratigraphic position he knew. Smith thus made use of the principle of superposition to enable him to determine the relative relationship of one fossiliferous bed to another and therefore one fossil aggregate to another. Once the relative relationship of fossil-bearing beds had been established, Smith could then use any given isolated fossil aggregate to determine the position of the rocks bearing it in an overall rock succession.

Richardson and other friends urged Smith to publish his views, warning him that someone else might publish similar ideas before him. Smith submitted a short series of notes containing his ideas to his friends but was always too short of funds to be able to devote himself, full-time, to writing a lucid account of his work for publication. Besides, his burning desire was to finish the geologic map. To that end he continued to travel, work, study, and compile data. A colored map such as Smith wanted to produce was expensive to publish, but with the aid of about 400 subscribers, headed by Sir Joseph Banks, and an enterprising map publisher, John Cary, the printing was begun in 1812. Smith worked closely with the printer, supervising the actual coloring of the map himself. Finally, in 1815, *The Geological Map of England and Wales* was ready. Proudly, on May 23, 1815, William Smith, Mineral Surveyor, attended a meeting of the Board of Agriculture with the first finished copy of his geologic map, and the first copies were delivered to the public August 1, 1815. Smith was at last able to lay the product of years of examination, comparison, and synthesis before the eyes of the world. His work was a magnificent achievement involving wide knowledge not only of rocks but also of the fossils in them. Its merit is not only in its magnitude but also in the fact that it was based on a thorough knowledge of the succession of faunas. Smith's greatest contribution to geology was the demonstration of the validity of the principle of faunal succession. Smith not only formulated the hypothesis but also demonstrated, through his geologic map, that it is a broadly applicable general truth— a principle.

Through the inquisitiveness of Hooke, the plaintive chiding of Cuvier, and the recognition of a particular sequence by the Abbé Giraud-Soulavie, the initial steps toward recognition of faunal succession as a valid principle had been taken. The final step was taken by Smith, who induced the principle.

Unhappily, Smith's great accomplishment was ignored at the time of its publication by most of those who might be considered geologists. His limited education and the fact that he earned his living by practical application of his knowledge apparently conspired to make him unworthy of notice by the learned gentlemen of his time. A much more widely used geologic map of Britain was that compiled by George Bellas Greenough. It was based on lithology and succession of rock units. Greenough's map was published in 1819 with financial support from the Geological Society of London, and went through several printings and revisions. Only about four hundred copies of Smith's map were published; most of these were bought by his friends and acquaintances. It is of interest to note that Greenough was a man of great wealth, a founder of the Geological Society of London and its first President, and as President of that Society, a member of the committee that ruled on the use of the Society's funds for financing publication of his geologic map. Greenough did note that Smith's map had been of help to him in compiling his map, and that as he had thought that Smith's map would probably never be published when he started work on his own, he had gone ahead with it.

Not until the latter part of his life did William Smith receive appropriate acknowledgment for his contribution to geology. Probably the high point in this recognition was Adam Sedgwick's address to the Geological Society of London, given in February 1831 when he presented the Society's first award of the Wollaston Medal to Smith. He stated in that address that there could be no doubt but that it was the Society's duty "to place our first honour on the brow of the Father of English Geology" (Sedgwick, 1831, p. 279). Sedgwick, who was then President of the Geological Society, went on to say:

> If, in the pride of our present strength, we were disposed to forget our origin, our very speech would bewray us; for we use the language which he taught us in the infancy of our science. If we, by our united efforts, are chiseling the ornaments, and slowly raising up the pinnacles of one of the temples of Nature, it was he that gave the plan, and laid the foundations, and erected a portion of the solid walls, by the unassisted labour of his hands. The men who have led the way in useful discoveries,

have ever held the first place of honour in the estimation of all who, in aftertimes, have understood their works, or trodden in their steps. It is upon this abiding principle that we have acted; and in awarding our first prize to Mr. Smith, we believe that we have done honour to our own body, and are sanctioned by the highest feelings which bind societies together.

William Smith provided geologists with a key by which the doors of past time might be unlocked: the principle of faunal succession led others to use specific fossil aggregates to delineate units in a time scale. A sequence of large magnitude fossil aggregates was established by geologists who used the principle of fossil succession as they worked with fossiliferous rocks early in the nineteenth century. Each aggregate was first obtained in an area in which its position in the overall sequence of such aggregates could be established, and in which there were a relatively large number of the fossils taken to constitute the diagnostic aggregate. Such an area may lend its name to the time unit based upon the faunal aggregate found in it. An aggregate may be unique to a particular span of past time because it is composed of many different species, some of which change through evolutionary development to new and different species. Aggregates change in time as the processes of evolution proceed. Because the processes of evolution do proceed and aggregates change, one aggregate is unique for only a relatively short span of time. Because it is, a time unit may be based upon it.

The principle of faunal (and floral) succession thus pointed to a series of unique fossil aggregates that could be recognized after fossils had been collected in precise superpositional order. Subsequent to Darwin's contribution of natural selection, it has come to be realized that organic evolution through natural selection lies behind the unique aspect of each of the aggregates and makes them the records of unique events in time.

The Setting

The period spanning the late eighteenth and early nineteenth centuries was a time of turmoil in geology. On one side, Werner and his students were insisting on their "neptunistic" views of the origin of the rocks of the earth's crust. This school of thought was being attacked most strongly by the vulcanists, who maintained that basalts and other volcanic rocks could not be precipitates from one universal ocean. At the same time, Cuvier's catastrophic history of the earth held a dominant position, and was accepted with slight modifications by all who took literally the Book of Genesis story of the earth's beginning. Opposed to the Cuvier catastrophic view was that of uniformitarianism, as outlined by Hutton and amplified by Playfair and Lyell.

The principle of uniformitarianism encompassed the vulcanist's observations, that volcanoes were formed by naturally occurring phenomena, which could be demonstrated to have occurred at several different times during the history of the earth. Hutton added to the arguments against Werner's opinions about the formation of basalts and granites. Of particular importance were Hutton's observations of granite veins cutting other rocks, including stratified layers. These observations led him to suggest that some granites were not chemical precipitates from an ocean, but that they had an igneous origin, and were molten when they forced their way into fissures and cracks in other rocks.

As the four ideas—neptunism, vulcanism, catastrophism, and uniformitarianism—were discussed by geologists around the time of the turn of the century, and evaluated in terms of the available evidence, some men accepted parts of one and parts of another. One of the most often used details of the neptunistic view was that certain rock units were the same age wherever they were found, and, hence, to establish steps in earth history, one needed only to work out the sequence of rock units.

Then, into the maelstrom of geologic discourse, came the principle of faunal succession. It provided the key to the establishment of major time units in the prehistoric past by permitting interpretation of the fossil record for the passage of time. Interpretation of the fossil record is requisite to recognition of any unit within the interpretive time scale so established.

The Transition from Descriptive to Interpretive Units

During the latter part of the eighteenth century and into the nineteenth, geologists in nearly every country in Europe were engaged in tracing strata, collecting fossils from them, and establishing local stratal sequences. At that time, the major divisions of the rocks of the earth's crust were considered to be those recognized by Lehmann, Werner, and Arduino. The terms Primitive, Transition, Secondary, and Tertiary were used most commonly. Individual rock bodies ("formations" in today's usage) were assigned to one of these major divisions according to their general appearance (highly indurated, coarse grained, highly folded and cleaved, richly fossiliferous and little indurated, etc.) and, to some extent, on their stratal superpositional relationships. The major divisions, Primitive, Transition, Secondary, and Tertiary (or in some uses just Primitive, Secondary, and Tertiary), were thought to have much the same significance that we today attach to eras although no interpretation of the faunas contained in them was made. Cuvier and Alexandre Brongniart's study of the Tertiary rocks in the Paris Basin is an example of the kind of stratigraphic studies that were carried out early in the nineteenth century. They used the chalk as their starting point and worked out the stratal order of the rock units throughout the Paris Basin, using both the lithological and paleontological characteristics of each. They described each rock unit in considerable detail, noting that each bore a characteristic fossil aggregate. Italian geologists similarly studied various deposits that could be placed in Arduino's Tertiary, and British geologists were tracing out the rock units and collecting fossils from them that were later to lead to their recognition as being Jurassic in age. The oldest-looking, most contorted and cleaved rocks were commonly classed as either Primitive or Transition.

The rock units now called "formation" or "group" were then thought to be subdivisions of the major divisions. Groups of formations were sometimes classified as "series"; such series were considered to be first-order subdivisions of the major divisions. As work with stratal units developed during the nineteenth century, formations came to be thought of as second-order subdivisions. No interpretation was needed to recognize any of these stratal units; that is, they were all descriptive rock units. The fossils in the rocks were used in the same manner as size of the grains forming the rock unit or its color—descriptively—to denote and map a unit. Fossils were not interpreted for their time significance until after William Smith demonstrated the principle of faunal succession and used it to put together his geologic map. Application of the principle of faunal succession demonstrated that some rock units (today's formations) that had been placed in the Primitive (in the sense of Werner) on the basis of general appearance turned out to belong with much younger rocks. Other rock units turned out, when their fossil content was analyzed, to be older than had been assumed on the basis of their appearance.

The rock units that we today consider formations and the groups and series of such formations were descriptive units having no place in a time scale based upon interpretation of fossils, but the names attached to many of these rock units were widely used by and well known to geologists before Smith demonstrated the validity of the principle of faunal succession. Because of this, many of these same names were used for interpretive units in the time scale as it was developed. Furthermore, many descriptive units were included without any real justification with the interpretive units in the time scale. For example, the Oswegan, Niagaran, and Cayugan were recognized in New York State as subdivisions of the Silurian System there. These units were constituted of several formations and so had been termed "series" in the descriptive sense. Because they had the designation "series," they were later included with interpretive units in the time scale and were thought to be time-stratigraphic series. But, in fact, the faunas from these bodies of strata had never been interpreted for their time significance. Indeed, both the Oswegan and Cayugan bore very few fossils, and those few were indicative only of special environmental conditions. By much the same manner of thinking, the evaporites of the "Ochoan Stage" in the southwestern United States have come to be considered as both a rock unit and a unit in the interpretive time scale.

Many of the widely used descriptive units did bear fossils that, when analyzed using the principle of faunal succession, proved to be a fossil aggregate diagnostic of a time unit in an interpretive time scale; thus many descriptive units became interpretive ones, and today bear the same names. Among the major units of the interpretive time scale that were originally descriptive rock units are the Cambrian, Carboniferous, Jurassic, Cretaceous, and Tertiary. Units that were based on interpretation of fossils from their inception are the Ordovician, Devonian, Permian, and the Tertiary Epochs.

CHAPTER SIX # Growth and Development of the Periods

With the background of thought, debate, and growth of principle discussed to this point in mind, let us proceed to the growth and development of the major time units, which most geologists use—the periods. In tracing the story of each of the periods, we will try to discover when it became a time unit based upon faunal analysis—an interpretive unit. Although there is no absolute agreement among geologists concerning which period names to use and the duration of certain periods, most of the periods discussed here form a sort of frame of reference used by geologists in all countries.

The Tertiary

In a letter to Professor Antonio Valisnieri at the University of Padua, Giovanni Arduino described certain mountains as a succession of hard limestones, consolidated and unconsolidated sand and gravel, and volcanic glass, and suggested that these mountains, and the rocks of which they were composed, should be termed "Tertiary" because they could be seen in superposition above the mountains (and rocks) he called "Secondary." Arduino

The Periods

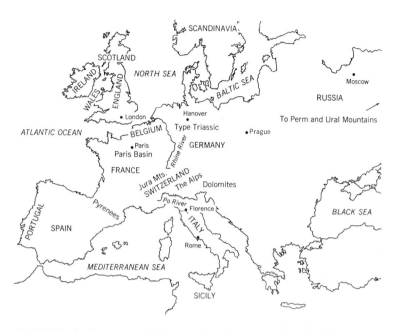

FIGURE 5. Outline map of Europe, showing some places of importance in the recognition of the major units in the time scale. [Adapted from A. O. Woodford, *Historical Geology*, W. H. Freeman and Company, San Francisco, © 1965.]

also noted that the rocks he classed as Tertiary were composed for the most part of shell fragments and rock particles derived from Secondary and Primitive rocks. Arduino judged that the deposition of the Tertiary rocks represented a major interval in the history of the earth's crust.

The Tertiary rocks were traced widely in northern Italy and many naturalists made large fossil collections from them. Rocks with similar aspect and relatively similar fossils were discovered in lands other than Italy. One of the most comprehensive descriptions of Tertiary rocks was Cuvier and Brongniart's study of the succession of Tertiary strata in the Paris Basin, made during the first decade of the nineteenth century. It was one of the most detailed descriptions of a stratal sequence published to that time.

The Tertiary became a time period based upon analysis of the faunas after Lyell and Gerard Deshayes studied the fossil mollusks from these rocks in the late 1820's and early 1830's. Lyell's publication in 1833 of the Tertiary subdivisions (the epochs) may be taken as the first published analysis of Tertiary faunas for their time significance. That year may be considered the date at which the Tertiary became an interpretive unit—a period.

The Carboniferous

Coal-bearing beds had been studied extensively for many years, because their economic importance was so great, before they were formally designated as a major, descriptive rock unit. The coal beds were recognized near the base of Werner's Flötz rocks, and in many localities were subdivided by the miners. Many details of particular coal-bearing beds were known by the first decade of the nineteenth century. In 1808, J. J. D'Omalius d'Halloy, working in Belgium, categorized the coal seams and related shales and sandstones as the upper part and the limestones beneath them as the lower part of the descriptive rock unit he labeled the *Terrain Bituminifere.*

A few years later, in 1822, W. D. Conybeare and W. Phillips summarized the existing knowledge of British strata from the youngest rocks to the base of the Old Red Sandstone in a monumental tome, *Outlines of the Geology of England and Wales.* Detailed and clear in presentation, this work remained for several years an outstanding basis from which British geologists could proceed in their stratigraphic inquiries.

Conybeare and Phillips discussed in it the *Carboniferous*

Order as one of the major divisions of the rocks they had under scrutiny. Within the "Carboniferous Order" they recognized four subdivisions: from youngest to oldest, the Coal Measures, Millstone Grit, Carboniferous or Mountain Limestone, and Old Red Sandstone. They stated (1822, p. 333),

> The class of rocks thus constituted will contain not only the great coal-deposit itself, but those of the limestone and sandstone also on which it reposes, which, though entitled to the character of distinct formations, are yet so intimately connected with the above, both geographically and geologically, that it is impossible to separate their consideration. . . . The epithet Carboniferous is of obvious application to this series.

They considered the rocks of the Carboniferous Order to be a unit distinct from either Werner's Transition or Flötz yet probably closer to the Transition. They felt the use of a new name rather than one of Werner's terms was justified because they wished to get away from the stigma attached to Werner's terms.

The Millstone Grit, Mountain Limestone, and Old Red Sandstone were included with the Coal Measures because Conybeare and Phillips considered all of these rocks to be closely associated in the districts where they were found. To them, the four considered together constituted a natural group of rocks. Fossils found in the Mountain Limestone were listed and noted to constitute a fauna that was different from any in overlying rock units.

Shortly after Conybeare and Phillips described the Carboniferous Order in England, d'Omalius d'Halloy described the *Terrain Houiller* as including only the coal-bearing layers, and he used the term *Terrain Anthraxifere* for beds under the coal, which he considered equivalent to the British Mountain Limestone and Old Red Sandstone. He stated that the primary characteristic of the *Terrain Anthraxifere* was the fossil contents, most of which were in the upper beds. D'Omalius d'Halloy's setting forth of two *terrains* where Conybeare and Phillips had delineated one, set the stage for future work in which two subdivisions of the Carboniferous have been widely recognized. The upper part commonly includes the Coal Measures, and the lower part, the Millstone Grit and the Mountain Limestone. The Old Red Sandstone could no longer be considered to be of Carboniferous age when it was shown to be the equivalent of a group of rocks bearing the fossils that were used as the basis upon which the Devonian Period was established in 1839.

Shortly after the Carboniferous was recognized as a separate group of rocks, its fossils were the subject of detailed study. John Phillips described the fossils from the Carboniferous limestone in Yorkshire in a report published in 1836. Shortly thereafter, Frederick McCoy's work on the Irish Carboniferous limestone fossils appeared. Laurent G. de Koninck spent nearly fifty years working on the faunas in Belgium and suggesting paleontological subdivisions of the Carboniferous limestones there. His monographic studies, together with those of Phillips and McCoy, are basic to paleontological research concerning fossils from the lower part of the Carboniferous System.

The fossils in the Carboniferous limestone rocks were marine invertebrate faunas. In the Coal Measures rocks, plant fossils were well known, having been noted by the miners and described in some detail by Brongniart (1821) and others. Of particular interest to the workers in and near the coal mines were the fossils of stems of plants and trunks of trees, some of which were seen in their original positions, standing upright with the rock layers compacted about them. These early floral and faunal studies and those that followed permitted recognition of the Carboniferous rocks wherever they were found and contributed toward making the Carboniferous a fully interpretive unit in the time scale.

The Cretaceous

In the same year that Conybeare and Phillips recognized the Carboniferous as an important rock unit, d'Omalius d'Halloy mapped and named the *Terrain Cretacé*, or Cretaceous System. Rocks grouped in this system had been studied for many years before its formal establishment, but most of the investigations included only small parts of the sequence. Werner and his students had traced limestones and sandstones in Saxony; they defined the position of these as at the top of the Flötz rocks. William Smith distinguished two kinds of chalk, a greensand, and a clay as formations. His upper chalk was easily traced in southern England, and it could be recognized in northern France. D'Omalius d'Halloy traced it from there into Belgium. Other geologists recognized it in Holland, Denmark, northern Germany, Poland, and Sweden.

D'Omalius d'Halloy had been engaged by the Baron de Montbret, who had charge of statistical information in France, to make a map showing the different kinds of rock masses that underlay that country. Since a number of geologic investigations had been made in France, D'Omalius d'Halloy had a large amount of information at his disposal for his map-making project. He had first to evaluate the work of others, then to portray the rock divisions that he deemed valid on the map. He concluded that the third of the five groups he recognized within the Secondary rocks should be called the *Terrain Cretacé*. This group included not only the chalk itself but also the greensand, ironsand, and marl beneath it. He stated in the text that accompanied the map (1822, p. 368–369): "The chalk formation, such as I have determined it in a preceding memoir, i.e. comprising the tuffas, sands, and marls, which occur beneath the true chalk, constitutes the third group." He demonstrated that the chalk formation as he recognized it was an extensive unit, known over much of the low

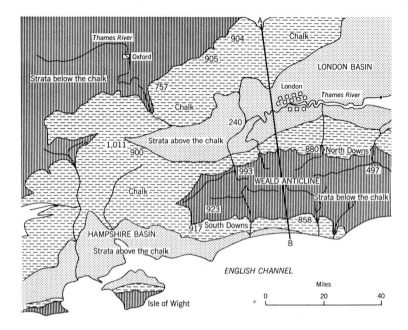

FIGURE 6. Simplified geologic map of part of southeastern England after William Smith (1815). The numbers are spot elevations in feet. [Adapted from A. O. Woodford, *Historical Geology*, W. H. Freeman and Company, San Francisco, © 1965.]

territory in the northwestern part of France. From the map, and from d'Omalius d'Halloy's discussions, the type area for the *Terrain Cretacé* can be regarded as the Paris Basin and those parts of Belgium and Holland adjacent to France in which the deposits crop out.

Conybeare and Phillips also used the word "Cretaceous" in 1822 but included some chalk beds and marls overlying the chalk beds with what d'Omalius d'Halloy had placed in his *Terrain Cretacé*. Conybeare and Phillips discussed two groups or series of rocks, the chalk and the beds that were beneath the chalk and

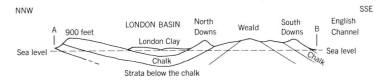

FIGURE 7. Approximately north-south section from north of London to the English Channel along Line A–B in Figure 6. Dip of the beds exaggerated to illustrate structure. [Adapted from A. O. Woodford, *Historical Geology*, W. H. Freeman and Company, San Francisco, © 1965.]

yet were above the Oolite Series, as the British equivalents of the rocks d'Omalius d'Halloy had included in the *Terrain Cretacé*. Similar subdivisions were recognized by other workers. Gideon A. Mantell, in 1822, published *The Fossils of the South Downs, or Illustrations of the Geology of Sussex*, in which he recognized two groups of rocks, as had Conybeare and Phillips, and went on to illustrate many of the fossils belonging to each formation he had recognized within the two major divisions. Perhaps of most interest in Mantell's work was the description of the fossil flora and reptiles from one of the lower formations, the Wealden.

During the 1830's, French geologists were particularly busy studying Cretaceous deposits. Auguste de Montmollin, working in the Jura Mountains, recognized a sequence of limestones and marls bearing fossils similar to those in the British greensands beneath the chalk and termed them the *"Terrain Cretacé du Jura."* Other workers recognized the same sequence in different districts, and O. Dufrenoy and J. Elie de Beaumont clarified the Cretaceous sequence by noting that it had two basic subdivisions, a "Lower Group," which included the Wealden formation, the "Neocomian Beds" newly recognized in the Jura, and the greensand and marls, and an "Upper Group," which was the white chalk. They thus amplified d'Omalius d'Halloy's original usage and pointed out that the two groups recognized by British geologists were indeed within the Cretaceous and formed two major subdivisions of it. Other French geologists studied the fossils and the rock layers from which they came with great diligence. One such study was that made by A. Leymerie, who worked out the Cretaceous sequence in the Aube area. He described the formations in detail and carefully noted the positions of all of his fossil assemblages. He pointed out similarities of the fossils from the White Chalk to those from similar beds in Britain. He also compared the fossils from the greensands and marls with fossils from beds of the same stratigraphic position in Britain, and pointed to formations in other parts of France bearing the same fossils. Study of the Cretaceous in France culminated during this period in the grand efforts of Alcide d'Orbigny, who in his *Paleontologie francaise*, published in 1840–1846, and in a text, *Cours elementaire de Paleontologie et de Geologie stratigraphiques*, published in 1849–1852, was able to divide the *Terrain Cretacé* into small units that he termed "stages." His stages were based upon faunal content and not upon mineralogic content because, d'Orbigny noted, mineral associations may occur at any place at any time. The stages were based upon complete rock sequences from which fossils could be collected in succession. The actual basis for each stage was a fossil aggregate that was unique to it. The stages were the most refined subdivisions of the Cretaceous at the time they were established and permitted widespread recognition of small parts of the system. These fossil studies led to establishment of the Cretaceous as a period founded on a unique fossil aggregate.

The British idea of a two-fold division of the Cretaceous found an expression in North America. R. T. Hill, working in Texas on a sequence of fossiliferous limestones and sandstones, denoted a "Gulf Series" as being essentially equivalent to Middle and Upper Cretaceous and a "Comanche Series" as Lower Cretaceous. These series terms were used as subdivisions of the Cretaceous System by Hill in 1887, but T. C. Chamberlain and R. D. Salisbury raised the "Comanche Series" to "Comanche System" in their textbook published in 1906. They restricted the name Cretaceous to the upper part of the Cretaceous System as it was recognized by Europeans. The Chamberlain and Salisbury argument was based on the physical evidence they saw in North America for the existence of two units of system rank instead of one. This proposal, unlike a similar one for the Mississippian and Pennsylvanian, did not become widely used or accepted.

The Jurassic

Jurassic rocks have attracted more attention from geologists than those of most, if not all, other systems. The Jurassic rocks crop out in much of the southern part of England and in France, Germany, and Switzerland. Study of the rocks now grouped in the Jurassic led William Smith to the data from which he induced the principle of faunal succession, Albert Oppel to the fossils and rocks he used to delimit the most refined possible subdivisions of the geologic time scale, Jules Marcou to the first demonstration of the existence of broad patterns of distribution of organisms in the prehistoric past similar to those in modern seas charted as zoogeographic provinces, and Armanz Gressly to the definitive demonstration of "facies," which is a designation for the differing lithologic and paleontologic aspect of rocks of the same geologic age seen in different localities. Detailed study of Jurassic rocks thus was the source of many of the guidelines and tools that a stratigrapher uses in interpreting the earth's history.

The name Jurassic had its beginning in the observations and descriptions made by the adventuresome, widely traveled scientist, Alexandre von Humboldt, during a geologic excursion he made in 1795 through southern France, western Switzerland, and northern Italy. During that trip, von Humboldt concluded that the *Jura-Kalkstein,* a massive limestone formation in the

Jura Mountains in Switzerland, which Werner had included in his Muschelkalk unit, was a distinct rock unit. He mentioned the *Jura-Kalkstein* in his book, *Ueber die unterirdischen Gasarten,* which was published in 1799. No fossils were described. The name *Jura-Kalkstein* as von Humboldt used it in 1799 referred to only a part of what is now known as the Jurassic System. Although the word "Jurassic" was not used, that report of von Humboldt's designating a Jura limestone formation is commonly cited as the first attempt to establish one of the major units that later became a period in the geologic time scale. The unit proposed by von Humboldt was a descriptive rock unit. Many years elapsed and many studies were made of the fossils from the rocks that were eventually included, before the Jurassic Period became an interpretive unit.

While von Humboldt was convincing himself that the *Jura-Kalkstein* was a distinct formation of considerable import, William Smith had been engaged in canal construction in the southwest of England. In this endeavor, he had daily come into intimate contact with rocks that have subsequently been identified as Jurassic. In the period from 1793 to 1798, Smith worked out a sequence of stratal units from the coal through the chalk. He noted both the lithologic aspects and the characteristic fossils of each. His observations were so detailed that later observers were able to use them with little or no emendation. Conybeare and Phillips recognized in their treatment of British strata an Oolite Series with two formations, the Oolite (above) and the Lias (below), as subdivisions. From the 1820's on, details of rocks now known to be Jurassic were assiduously examined by British geologists and naturalists. Fossils were collected and, in many cases, their exact position in stratal layers was noted. Studies on both sides of the English Channel revealed that many of the details of the British Oolite and Lias could be distinguished on the Continent as well. Indeed, Dufrenoy and Elie de Beaumont used many of the British formation names in the text for their geologic map of France, published in 1841 and 1848.

As study of the Jurassic progressed, it attracted the attention of German geologists and paleontologists, and several studies were published which demonstrated, through comparison of fossils, that the British Oolite and Lias had equivalent units in northern Germany. Then Leopold von Buch, who had examined some of the Alpine Jurassic with von Humboldt in 1795, turned geologists' attention to rocks in southern Germany with his study, *Über den Jura in Deutschland,* published in 1839. In that work, he divided the equivalents of the British Oolite and Lias that he observed in southern Germany into three units: Upper or White Jura, Middle or Brown Jura, and Lower or Black Jura. He noted that the lowest unit was widespread and that the upper two formed massive beds above it. Through the study of fossils, the units were compared with those of the same age in France and England. Von Buch also noted that the three basic groups might be further divided by using fossils. This suggestion was later followed by F. A. Quenstedt and his pupil Albert Oppel. Quenstedt popularized his published work by including many attractive pictures; he taught the local farmers in various areas in southern Germany to recognize certain fossil species and to collect any fossils they found, noting precisely where each had been obtained. Some of the farmers were even trained to identify certain characteristic aggregates.

Today, many Jurassic deposits and their fossils are known in great detail. Knowledge of the Jurassic faunas and strata reached its culmination in the work of a twentieth-century geologist, W. J. Arkell, who spent a lifetime examining these faunas and strata. Arkell's magnificent contributions include two large volumes, *The Jurassic System in Great Britain* and *Jurassic Geology of the World.* Probably no other person in the history of geology has known so much about the history of the earth during a single period.

Such has been the history of the Jurassic. Initially a descriptive (rock) unit, it became an interpretive unit in the time scale based upon faunal analysis through scores of studies of the faunas from Jurassic rocks. Widely studied, it has been the proving ground for many important stratigraphic principles. One may point to the stratal and faunal details set forth by William Smith and the amplification of Smith's work by Conybeare and Phillips as the real starting points leading toward recognition of the Jurassic Period as a time unit based upon a distinctive faunal aggregate. One must also note the significant contributions of von Buch. It was he who first included within the Jurassic all the rocks and faunas now commonly considered to be Jurassic in age. He also suggested possibilities for further researches that permitted division of the Jurassic into smaller time units.

The Quaternary

The rocks now known to be Cretaceous and Jurassic in age attracted much attention from naturalists because they were widely encountered, easily examined, and they yielded good fossils. During the latter part of the eighteenth century, many English clergymen and schoolmasters spent their leisure hours collecting fossils from accessible exposures. At the same time, naturalists in France and Italy were similarly engaged. They, however, were collecting from rocks that were realized to be equivalents of Arduino's Tertiary rocks. Arduino noted in his letter to Professor Valisnieri in 1760 that certain rocks that he categorized as volcanic lay above the Tertiary rocks. Rocks that were demonstrated to be equivalents of Arduino's volcanic rocks were also examined by naturalists and fossils were collected from them.

The status of these rocks had been considered by a number of geologists early in the eighteenth century. The English commonly followed the decisions made by Professor Buckland, who, in 1823, gave the name *Alluvium* to all deposits forming at present and the name *Diluvium* to the deposits beneath those of the Alluvium but above the Tertiary rocks. The deposits of the Diluvium were considered by Buckland to have formed during the Noachian Deluge. In Werner's classification, which was followed by many European geologists at that time, Tertiary rocks and any deposits above them were placed in the *Aufgeschwemmte-Gebirge*.

Paul G. Desnoyers, in discussing deposits in the basin of the Seine in 1829, classified under the name Quaternary, certain marine, lacustrine, alluvial, and volcanic materials that he thought to be younger than the Tertiary rocks. He was of the opinion that these deposits constituted a group distinct from the Tertiary strata and that it was an important group of deposits in the succession of rocks in the earth's crust. Desnoyers' Quaternary was thus a descriptive group of rocks and some unconsolidated material; it had few faunal remains by which it could be recognized outside of its typical area.

H. P. I. Reboul discussed the Quaternary Period in 1833 and pointed out that it was typified by fossils of animals and plants that were like organisms now living. He considered that the Quaternary had two subdivisions, historic time and prehistoric post-Tertiary time. The Quaternary was distinguished from the Tertiary by its fossils, which were remains of organisms like those now living in the area in which the fossils were found. Tertiary fossils, by contrast, were remains of types of organisms no longer found in the same general area. Reboul considered that the deposits to which Desnoyers had given the name Quaternary were actually Tertiary. C. A. von Morlot, in 1856, followed the Reboul usage of "Quaternary" but recognized glacial sequences as divisions of it. He, too, included historic time within the Quaternary. Most geologists have followed this usage, with, from the geologic viewpoint, glacial deposits attracting prominent attention. Indeed, one entire area within geology is concerned with glacial activity and landforms of glacial and modern time.

The Triassic

The lowest part of the *Terrain Secondaire,* or Werner's *Flötz-Gebirge,* attracted little attention from paleontologists and stratigraphers for some time after they had worked on the more easily accessible and more richly fossiliferous upper parts of that terrain. True, Lehmann and Füchsel had studied and described in detail formations now included in the lowest part of the *Terrain Secondaire,* the Triassic, but a comprehensive recognition of it as a system awaited the researches of several German geologists during the early part of the 1800's.

In the course of early studies, local details of three rock units, the Bunter Sandstone, Muschelkalk Limestone, and the Keuper Marls and Clays were worked out. Fossils from the Muschelkalk in Thuringia, the study area of Füchsel and Lehmann, were well illustrated and described by E. F. von Schlotheim in 1823 in *Nachtrage zur Petrefaktenkunde.* Others, such as Peter Merian in 1821, F. L. Hausmann in 1824, and F. Hoffman in 1823, established the superpositional relationships of the Bunter Sandstone beneath the Muschelkalk Limestone in the Black Forest area and in northern Germany. The first use of the name Keuper for the clays and marls above the Muschelkalk was put forward in the work of H. von Dechen, C. von Oeynhausen, and C. La Roche following their trip through the upper part of the Rhine River area in 1825.

Friedrich August von Alberti began an intensive examination of the rocks that constituted the salt deposits in Germany and the rocks related to them when he went to work at the salt works

The Periods

in the town of Sulz in 1815. He studied the Bunter, Muschelkalk, and Keuper strata and their fossils closely and stated (1834, p. 323-324):

> Whoever examines more closely the foregoing analysis and tabulates all the fossils of the three hitherto separate formations; whoever examines, further, the transition of the different forms one into the other, and, indeed, considers the entire structure of the mountains and the markedly different character of the fossils of the Zechstein (Permian) from those of the Lias (Jurassic), will realize that the Bunter sandstone, Muschelkalk and Keuper are the result of a single period, their fossils, to use Elie de Beaumont's words, being the thermometer of a geological period; that their separation to three formations is not appropriate, and that it is more in accord with the concept of a formation to unit them into a single formation, which I shall provisionally name *Trias*.

Alberti was using the German meaning of "formation"—a "period" in today's terminology.

Following his analysis of the fossils, Alberti thus established the Triassic as an interpretive division in the time scale. It was founded on a distinctive fossil aggregate with relative time significance.

Alberti enumerated the lithological details of the several beds he included in the Bunter, Muschelkalk, and Keuper, and listed the fossils he found in each. His description included, at the top of the Keuper, sandstones that were later named Rhaetic. Subsequently, some geologists have asserted that all or a part of the Rhaetic Sandstones belong in the Jurassic.

As Alberti began with observations in western Germany, that area (see Fig. 5) may be considered the type for the Triassic. His three subdivisions, Bunter, Muschelkalk, and Keuper, remain applicable, with only slight modification, to primarily nonmarine beds of Triassic age in Europe. Geologists had to turn to the

Alps to find Triassic age beds deposited under marine conditions.

In the years following Alberti's recognition of the Trias as a period, geologists, primarily German and Austrian, examined Alpine areas to unravel the complicated geologic structures in that delightful though rugged terrain. Using established rock and faunal units, they were able, slowly, to gain some knowledge of the geologic intricacies and stratigraphic succession. Gradually, fossiliferous marine shales, limestones, and dolomites were recognized as correlative with the Bunter, Muschelkalk, and Keuper. The studies made by E. Mojsisovics of Triassic ammonites from the Alps and the Himalayas are basic works in the investigation of Triassic marine faunas. The names W. Waagen, C. Diener, and C. W. von Gumbel should be placed with Mojsisovics in the front rank of paleontologists who contributed to the understanding of Triassic marine faunas.

Alberti not only established the Triassic as a distinct time interval, but also, through his careful study of the fossils, provided paleontologists with data for uncovering the details of Triassic stratal relationships and faunal relationships. So fossiliferous were Triassic beds in the Alps and in the Himalayas, that a sequence of refined subdivisions have been delimited. Triassic strata, though not as accessible or widely distributed, have proved as fruitful for refined stratigraphic work as the strata of younger systems.

The Silurian and the Cambrian

The subdivision of Secondary rocks and those above them proceeded swiftly in the early part of the 1800's, but the Transition beds, or "old graywackes," beneath them remained essentially uninvestigated. To solve the stratigraphic puzzle posed by these heaps of contorted rocks was the task set for themselves by two friends in the Geological Society of London: Roderick Murchison and Adam Sedgwick.

Murchison was an enthusiastic geologic observer who, by virtue of his independent means, was able to devote his full energies to geologic inquiry. He had been an Army officer and a gentleman fox-hunter before he was convinced by Sir Humphrey Davy to sell his horses and hounds and to turn his energies to geology. During the winter of 1824, Murchison took courses in the sciences and joined the Geological Society of London, where he became acquainted with the current leaders of English geology, including Adam Sedgwick. Professor William Buckland of Oxford was particularly friendly, taking Murchison on geologic excursions that whetted his appetite for field work. Murchison then began to travel widely, searching for geologic data. He spent a part of the summer of 1825 in Devon and Cornwall, in the southwest of England. He went with Sedgwick to the Scottish highlands in 1827. He toured parts of France and Italy with Lyell in 1828, and then went on alone to visit the Alps. The next year he returned to the Alps with Sedgwick, and they also went to various areas in Germany. Murchison returned to Germany in 1830. His wife accompanied him on his excursions, patiently collecting fossils and making sketches while her husband rambled over hill and dale.

During the winter of 1830–1831, Murchison became determined to set foot on the geologic desert known at that time as the Transition rocks. Nearly all of the exciting stratigraphic work had been done on the Secondary rocks, and the textbooks of the time dismissed the older beds with a few lines. True, fossils had been recovered from the older rocks not only in Wales and the Lake Country in Britain, but also in Scandinavia, in eastern North America, and the Rhineland. No clue was known, however, of the stratigraphic relationships of the beds bearing them. The rocks from which these fossils came were so contorted that there seemed little hope of unraveling a stratal sequence. Murchison and Sedgwick had investigated the geologic complexities of the Scottish highlands and of the Alps—a valuable preparation for their task of making some sense from the crumpled pages of earth history that lay before them.

Adam Sedgwick, son of a Yorkshire clergyman, studied at Trinity College, Cambridge, and although he had not been trained in geology, was appointed Woodwardian Professor of Geology at Cambridge in 1818, at the age of thirty-three. He turned all his attention and energies toward geology after his appointment, and soon rose to a position of leadership among geologists of his day. He had keen powers of observation, and was willing to undergo extremes of physical exertion to learn about the phenomena of nature. At first, his reading led him to a Wernerian view of geology, but his own observations gradually turned him from that outlook. He traveled extensively in the British Isles, bringing back copious notes and a multitude of specimens for his museum. He joined the Geological Society and was quickly recognized as one of its leaders. His contact with Murchison there led them to join forces for excursions and finally for a joint attack on the complicated structure of the Transition rocks. Sedgwick, with characteristic verve, chose to study the most complicated geologic structures in rugged territory. He had begun his labors on the older rocks in the Lake District as early as 1822, working out the sequence of strata by examining lithologic characteristics and by relating rocks having similar mineralogic aspects. Then, in the summer of 1831, he went to work in North Wales and, during that and succeeding summers, was able to determine general stratal relationships. He used lithologic aspects of rocks to equate them to others of the same appearance and in this way built up a stratal sequence. He was able to trace contortions of the strata and elucidate the geologic structures, but his stratigraphic work included no descriptions of general characteristics by which the formations might be recognized outside of the study area. The formations bore fossils, but these had not been noted in his descriptions of the formations. Sedgwick had plunged into geologically unknown territory where relationships to any part of the then-known rock sequence were unknown. He worked out a stratigraphic succession valid in his study area. He was more interested in geologic structures than in fossils, and apparently gave little thought to their use, at least in the early part of his mapping.

Murchison began his investigation of the Transition beds in what may be considered a more orderly scientific manner. He left London by carriage with his wife and maid and traveled slowly westward, stopping at Oxford to see his friend, Professor Buckland. Buckland told him all he knew of the rocks that lay beneath the Old Red Sandstone (these red sandstones were widely recognized in South Wales and the Welsh Borderland: they had been placed at the base of the Carboniferous Order by Conybeare and Phillips) and noted that a good sequence could be seen on the banks of the Wye near the town of Builth. Murchison next visited the Reverend Conybeare and gleaned from him all he could. Then, he proceeded to visit all of the local amateur geologists and fossil collectors in the Welsh borderland country that he could find. They told him what they knew and showed him their collections. Of particular import were the observations and collections of the Reverend T. T. Lewis of Aymestry, who had determined the stratigraphic sequence of the rocks in his area and had recognized the fossils characteristic of each stratum. At last Murchison turned to his own field work in the hills of south Wales, near the town of Llandeilo. He worked back and forth across the Welsh borderlands, found a sequence of strata in the upper part of the Transition beds and traced it up to its contact with the Old Red Sandstone. He found plenty of fossils in the layers he examined and collected them carefully, noting their exact stratigraphic position.

Both Murchison and Sedgwick continued independently labors

in the "old graywackes" for the next three summers (1832, 1833, 1834) before they were prevailed upon to name the rock units they were investigating. Murchison had presented reports of his investigations to the Geological Society of London in September, 1831, in March and April of 1833, and twice in January, 1834. In the last report, he described "four fossiliferous formations" in detail and showed in tabular form their stratigraphic order, thicknesses, subdivisions, localities, and characteristic fossils.

Sedgwick, by that time, had presented two brief reports. The two men had spent a few days together in 1834 going over their respective areas. At that time, both were convinced that the upper units of Sedgwick's succession "plunged under" Murchison's lowest units and thus were definitely older.

In July, 1835, in an article in the *London and Edinburgh Philosophical Magazine,* Murchison named his rocks the Silurian System, after an ancient tribe, the Silures, that had inhabited the Welsh Borderland. In that article, he outlined two major subdivisions of the system, the Upper Silurian, which included the Ludlow Rocks and the Wenlock Limestone, and the Lower Silurian, which comprised the Caradoc Sandstones and the Llandeilo Flagstones. The deposits of the system were described as lying below those of the Old Red Sandstone and above the slates of Wales on which Sedgwick was working.

In August of 1835, Sedgwick and Murchison jointly presented a paper entitled, *On the Silurian and Cambrian Systems, exhibiting the order in which the older sedimentary strata succeed each other in England and Wales,* before the British Association for the Advancement of Science. Murchison affirmed the fact that the Ludlow, Wenlock, Caradoc, and Llandeilo Formations, each of which contained characteristic fossils, constituted one natural geologic system, the Silurian. He considered that the term Transition under which those rocks had formerly been grouped should no longer be used. Murchison noted that the upper beds

FIGURE 8. A sketch of Ludlow Castle. Exposures of Upper Ludlow Rock are shown on both banks of the River Teme, which flows past the castle. [Adapted from R. I. Murchison, *The Silurian System*, John Murray, 1839.]

of the Silurian System lay stratigraphically beneath those of the Old Red Sandstone and that the lowest beds, particularly where he had studied them in South Wales, were directly underlain by slates, which were included in the system being delineated by Sedgwick.

Sedgwick noted that the rocks he had been studying were not as fossiliferous as those of Murchison's Silurian System and that sequences of formations were difficult to discern. He named his rocks the Cambrian System, after the Roman name for Wales, and recognized within it three major subdivisions. To the stratigraphically highest, he gave the name Upper Cambrian group, noting that it connected with the Llandeilo Flagstones of the Silurian System and was widely exposed in South Wales. Beneath the Upper Cambrian group lay that of the Middle Cambrian, and beneath it, the Lower Cambrian. Slates, volcanic rocks, and graywackes predominated in these groups that together constituted the Cambrian System.

The two friends jointly offered an answer to the riddle of the

Transition beds and considered because they had subdivided the Transition rocks, that the term "Transition" need no longer be used. Explicit in Murchison's work was the fact that the formations he included in the Silurian System contained characteristic fossils by which they could be recognized in other parts of England and any other place. He had used the principle demonstrated by William Smith, that of faunal succession, in delimiting the Silurian Period. Sedgwick, on the other hand, had not utilized fossils and had not pointed out any means by which the Cambrian System could be recognized in other places than Wales. This omission led him and his students into one of the most bitter and prolonged controversies in the history of geology.

Murchison worked steadily on the Silurian rocks and faunas and published a comprehensive account of them in a book, *The Silurian System,* in 1839. Sedgwick continued to describe details of the Cambrian rocks, but did not describe any fossils. Other geologists began to make geologic maps of Wales, and they soon traced the lower part of Murchison's Silurian into the upper part of Sedgwick's Cambrian. The problem was compounded by the fact that any fossils found in the Cambrian were said to be Silurian fossils: because the Cambrian fossils found during this period were not analyzed carefully enough, the formations containing them were placed in the Silurian System. Because Sedgwick had failed to describe a characteristic Cambrian fauna, Murchison was free to identify all fossils found in the rocks he was trying to subdivide as Silurian. By 1842, Murchison maintained that the Lower Silurian contained the oldest fossiliferous rocks, and that fossils from Sedgwick's Cambrian System in North Wales were no different from those of the Lower Silurian. He continued to hold this view for the remainder of his life, and because of his personal popularity and the fact that he became Director of the Geological Survey of Great Britain in 1855 and remained in that position until his death in 1871, his opinion became that of the majority of English geologists and those in other countries. From the position expressed in 1835 that the Transition rocks should be reclassified as two systems of rocks, Murchison had moved by the 1850's to a use of "Silurian" that was synonymous with the "Transition" he had started out to divide. By that time he had reduced Sedgwick's Cambrian to a group of rocks he thought were Silurian in age.

Sedgwick argued bravely for the Cambrian System but was unable to meet the challenge hurled at him by Murchison (1852, p. 176): "was the Cambrian system ever so defined, that a competent observer going into an uninvestigated country could determine whether it existed there?" Murchison had defined the Silurian as a group of rocks bearing a characteristic fossil aggregate and had gone on to recognize these fossils in many places on the Continent. This, to him, was the proof of the existence of a Silurian Period, and where was Sedgwick's similar proof? Of course none could exist for Murchison had encompassed all fossiliferous rocks beneath the Old Red Sandstone in his expanded version of the Silurian. Not until the 1850's did Frederick McCoy and John W. Salter describe fossils from the rock units Sedgwick had classified as Cambrian.

Once Cambrian faunas had been described, they were recognized in other countries. Consideration of the faunas and the fact that they were widely known throughout Europe and North America led to the acceptance of the Cambrian as a separate period. Not until Charles Lapworth's discussion in 1879, however, did the Cambrian become fully reinstated as a period by British geologists.

The Cambrian Period, as Sedgwick thought of it, was a descriptive unit and so could not be recognized outside of its type area. Sedgwick's failure to realize that periods are not simply descriptive units, but are interpretive units based on analysis of faunas made him one of the tragic figures on the geologic stage.

Many lesser geologists have suffered a similar fate for the same reason.

The Devonian

After their joint presentation concerning the rocks beneath the Old Red Sandstone, Sedgwick and Murchison spent a part of the next summer and several other periods during the years 1836 through 1839 together, examining the contorted rocks in Devonshire. They remained close friends during their work together there and their European excursions. (Later, in the 1840's, however, a schism began to form after British Survey geologists realized that Murchison's Silurian and Sedgwick's Cambrian overlapped.)

Their attention had been drawn to these rocks by Sir Henry de la Beche, who had sent some fossil plants collected from beds he thought to be "old graywacke" (possibly Cambrian or Silurian in age) in that area. Murchison was intrigued by these for he had not found any plants among the Silurian fossils, and urged Sedgwick to join him in determining precisely from what position in the stratigraphic sequence the plants had come. Murchison thought the plants resembled those from the Coal Measures, and after he and Sedgwick had examined the outcrops in North Devon, concluded that the plant fossils had not come from the graywackes but from beds above, which could be identified as belonging to the Coal Measures. Sedgwick and Murchison thus established on their 1836 excursion that the plant-bearing beds belonged in the Carboniferous System; they remained interested in the contorted rocks beneath the plant-bearing beds.

Sedgwick, sometimes alone and sometimes with Murchison, returned on other trips to Devonshire in the hope of ascertaining the proper stratigraphic relationships of the contorted rocks there. The Old Red Sandstone, which had been widely identified overlying the Silurian, could not be found in that part of the country as it had been elsewhere. The contorted rocks, commonly labeled graywacke, resembled some of the rocks in North Wales and so the two investigators thought of them as possibly Cambrian judging from lithologic appearance. One difference between the rocks in Devonshire and those in Wales was that the Devonshire rocks included many limestone beds that bore fossils, many of which had been collected by local naturalists.

Murchison could see few resemblances between these fossils and those from the Silurian. William Lonsdale, a retired army man, had spent several years collecting fossils and had become a specialist on corals from the Carboniferous limestone. He had also worked with corals from Murchison's Silurian System before he began to examine coral collections from the contorted rocks exposed in South Devon. He concluded that the Devon rocks containing corals were intermediate in age between those of the Silurian and the Carboniferous systems: Although some of the Devon species were similar to those from the Silurian and others resembled those from the Carboniferous, the total aspect of the coral fauna was intermediate (in terms of its stage of evolution) between that of the Silurian and that of the Carboniferous. The rocks bearing the corals must therefore be equivalent to the red rocks termed Old Red Sandstone which lay between the Silurian and Carboniferous rocks in other parts of England.

Lonsdale's suggestion concerning the stage of evolution of the corals was made in 1837, but scarcely considered by Murchison and Sedgwick at first. They gathered more stratigraphic data and more fossils, which were turned over to Lonsdale for study. He continued to proclaim them to be older than those from the Carboniferous and younger than those from the Silurian. At last his evaluation was taken seriously by Sedgwick and Murchison and, primarily on the faunal evidence, but supported by their own observations that the rocks from which the fossils came underlay Carboniferous rocks, they recognized a new period. The first publication of the new division of the geologic succession was in the year 1839: Sedgwick and Murchison jointly proposed the name Devonian System to encompass the slates, sandstones, and limestones lying beneath the Carboniferous beds and bearing fossils intermediate in characteristics between those of the Silurian and the Carboniferous. They noted that the sequence of strata was clearest in North Devon where it was more fossiliferous and less contorted; but that it could also be seen in South Devon and that Lonsdale had first studied the fossils of the system there. The Devonian System in its type area was considered to be equivalent to the Old Red Sandstone, which was known at that time from many localities in South Wales and the Welsh Borderland.

The Devonian Period was established by using a unique fauna and the superpositional relationships of the rocks bearing that fauna. Rocks found elsewhere bearing the same faunal aggregate as those in Devon could be interpreted to be Devonian. To determine if such fossils could be found on the Continent in rocks between those bearing Silurian and Carboniferous faunal aggregates, Murchison and Sedgwick spent the summer of 1839 making an excursion through the Rhineland, the Westphalian region, and the Eifel district. Fossil collecting was of prime importance, and several collections were shipped back to London where they were examined during the following winter by Lonsdale, George Sowerby, and John Phillips, who was at that time preparing a study of the Devon faunas. Their painstaking comparisons disclosed that rocks bearing fossils similar to those of the Devonian System were present on the Continent, and that these rocks lay beneath those with Carboniferous faunas. This investigation with its careful paleontological analysis firmly established the Devonian as a distinct period based on a unique and recognizable fauna. John Phillips made a major contribution to the knowledge of the Devonian fauna with his study, *Figures and descriptions of the Palaeozoic fossils of Cornwall, Devon, and West Somerset; observed in the course of the Ordnance geological survey of that district,* published in 1841.

Following Murchison and Sedgwick's excursion into the Rhineland and the Eifel area, European geologists began to study intensively the rocks and fossils of that region. The rocks were described by Ferdinand Roemer, André Dumont, and Jules Gosselet, and the faunas by Guido and Fridolin von Sandberg in the years 1850–1856. Other paleontological monographs followed, and today, the Rhenish and Eifelian faunas form the bases for widely used subdivisions of the Devonian.

The Permian

Murchison went to Paris in the spring of 1840 to present some conclusions derived from the search for continental Devonian fossils that he and Sedgwick had made during the previous summer, and while there he learned of the geologic treasures locked within the vast territorial borders of Russia. He heard that fossils abounded, that the rocks were nearly flat-lying and little deformed. He longed to go and see for himself. If the succession of faunas characteristic of each period were the same as had been determined in Britain and seen on the Continent, it would argue that they pertained to all of Europe, and perhaps to the whole world.

Murchison and the French paleontologist Edouard de Verneuil arranged to make a geologic excursion in Russia during the summer. Murchison hurried home from Paris, packed, and left London with de Verneuil in May, 1840. They went first to Berlin where Murchison learned as much as he could about the geology of Russia from Humboldt and von Buch and others. From discussions with von Buch, he was able to decide on some localities to visit.

At the outset of the 1840 excursion, Murchison and de Verneuil were accompanied by Count von Keyserling and Professor Blasius. Then Lieutenant Koksharof, an army officer trained in geology at the Imperial School of Mines, joined them. The party spent the summer traveling along the northwestern border of the Russian platform, visiting the shores of the White Sea, stopping at Archangell, touring southward along the Dwina River into the edges of what is now recognized as the Permian basin area, then turning west and south to Moscow by way of the valley of the Volga. Murchison saw a most interesting sequence of Silurian, Old Red Sandstone, and Carboniferous rocks, flat-lying and only slightly consolidated. The deposits were fossiliferous, but their lithologic aspects were so different from those in England, that only by recognizing characteristic fossil aggregates could their relationships to the established periods be determined. Of particular importance was the find of fossil fish, similar to those in the Old Red Sandstone, together with shells of marine invertebrates that were well established as Devonian forms. Here was evidence for the correlation Lonsdale had made of the Old Red Sandstone with Devonian shelly fossil-bearing strata.

The following summer, Murchison and de Verneuil returned to Russia. Early in the visit, the two geologists attended festivities celebrating the marriage of the Czar's eldest son, and there became well acquainted with the Czar.

Murchison's contact with the Czar resulted in a more extensive excursion than the first, made with the same Russian companions; and he was permitted to examine all the records and collections of the Russian geologists. The travelers went from Moscow east across the Russian platform to Perm and the Ural Mountains. There, they turned south and traveled along the Urals for some distance before proceeding southwest to the Sea of Azov and then north back to Moscow. This trip took five months, and Murchison examined rocks, made fossil collections, and studied the records and collections of local geologists all along the way. Rock units were sketched on maps and geologic structures were worked out in a general way. Murchison recognized rocks belonging to the Silurian, Devonian, and Carboniferous systems by the fossils they contained. He reaffirmed the principle of faunal succession and demonstrated the widespread applicability of the periods.

In addition, Murchison established a new period. Before leaving Russia in the fall of 1841, he presented the Czar with a report of his knowledge of the geology of Russia west of the Urals and a geologic map. He also wrote a letter to the Academy of Sciences at Moscow proposing the name Permian, after the small town of Perm on the western flank of the Urals, for the new period. Upon returning to England, Murchison established the name in an article published in the *Philosophical Magazine* late in 1841. He noted that the Carboniferous rocks on the west side of the Urals were overlain by a sequence of marls, limestones, sandstones, and conglomerates that bore fossil brachiopods somewhat similar to those in the Carboniferous rocks. These overlying rocks also contained, however, fossil fish and amphibians similar to those found in the German Zechstein beds, which had the same stratigraphic position as the Magnesian Limestone in England. The Magnesian Limestone had been demonstrated to overlie the Carboniferous. Further, the fossil flora from the rocks he proposed to place in the Permian System appeared to be intermediate in aspect between the floras of the Coal Measures and the Triassic. From the unique characteristics of the flora and fauna, Murchison concluded that the rocks bearing these fossils should be included in a new period. Fossils and position stratigraphically above the Carboniferous were the bases for establishing the Permian Period. Murchison pointed out that the Permian System in Russia had as its equivalents in Germany the red sandstones called "lower new red" or *Rotliegendes,* and the lime-rich beds called Zechstein.

a: Carboniferous limestone
b: Goniatite flagstone
c: Sandstones, limestones, gypsum, and grits (Permian rocks part of)
d: Red sands and copper ores
e: Conglomerate and sandstone

FIGURE 9. Diagram used by Murchison (1854) to illustrate the stratigraphic superposition and continuity of strata bearing fossils indicative of the Permian (c, d, e) with those bearing fossils indicative of the Carboniferous (a, b). The stratigraphic section shows Carboniferous age limestones in the Gurmaya Hills, which are on the southwest flank of the southern part of the Ural Mountains near the village of Kundrofka on the Sakmarka River, overlain by the Goniatite-bearing flagstones also of Carboniferous age. These rock units are in turn overlain conformably by limestone and interbedded gypsum and sandstones that are Permian in age. [Adapted from R. I. Murchison, *Siluria,* John Murray, 1854. The drawing originally appeared in R. I. Murchison, et al., *The Geology of Russia in Europe and the Ural Mountains,* vol. I, John Murray, 1845.]

Murchison spent much of his time during the next three years preparing a monograph, *The geology of Russia in Europe and the Ural Mountains.* This tremendous study included a detailed account of the Permian System, a colored geologic map, and many other items of geologic interest concerning the area he had traversed. Its magnitude was made possible by his synthesis of many observations collected from Russian geologists with his own.

Geologic investigations west of the Mississippi River in the United States permitted Major Frederick Hawn to recognize Permian fossils from limestones and shales in the Kansas Territory in 1858. Benjamin Franklin Shumard described Permian fossils from the now well-known limestones exposed in the Guadalupe Mountains in west Texas in the same year. Subsequent study of the American Permian deposits has revealed extensive sequences of fossiliferous strata in west and central Texas and in the part of the western United States known as the Great Basin. In the period from 1840 to 1890 a number of studies of the European and Asian strata made both nonmarine and marine Permian fossils and rocks well known. H. B. Geinitz and A. von Gutbier described the fossils of the German Permian, and W. King, those of the Magnesian Limestone in England. A. Karpinsky added a lower unit, the Artinsk Stage, to the Permian System originally outlined by Murchison, and it was widely traced by Russian geologists from the Arctic to the Caspian Sea. G. G. Gemmellaro studied the marine Permian faunas from Sicily. W. Waagen, in a series of papers, made the faunas of the Productus Limestone in the Salt Ranges of the Punjab district in India well known, and J. Wanner reported on the faunas from Permian Limestones on Timor Island in Indonesia. These studies enriched knowledge of the Permian faunas and demonstrated beyond doubt that the period was a distinct major time interval.

The Ordovician

About 1850 trilobites began to be described not only from Sedgwick's Cambrian rocks but also from rocks that lay beneath those bearing fossils confirmed to be Silurian and above those of the Cambrian on the Continent as well as in England. Joachim Barrande, a Frenchman exiled with the royal family in 1820, spent the greater part of his life in Prague, serving as tutor and administrator to Prince Henry of Chambord. Barrande devoted much study to the rocks and fossils in the area around Prague. That area is now known as the Barrandium. Fossil collecting there was easy, and the fossils were exquisitely preserved. Soon, Barrande had a vast collection and, in a series of 22 large volumes, published from 1852 until the year of his death, 1883, he described and illustrated the fossils in his collections. He noted that most of them were Silurian, and titled his work *Systeme Silurien du centre de la Boheme*. Using his knowledge of the superposition of faunas, Barrande designated eight *etages* labeled A through H. *Etage* D contained the typical Lower Silurian fauna that Murchison had recognized in the Llandeilo and Caradoc stratal units. *Etage* E had the fauna that Murchison had described from the Wenlock Limestone. The stratigraphically higher *etages,* F through H, were considered equivalent to the Upper Silurian of Murchison. Beneath *etage* D, the rocks were primarily schists and graywackes. *Etage* C did have trilobites that Barrande termed a "Primordial" fauna. Similar superpositional relationships between those strata with Silurian faunas and those beneath containing trilobites were described from Norway and Sweden in a series of papers published in the 1850's.

Geologic investigation in North America had been given considerable impetus by the establishment of the New York State Geological Survey in 1836. In reports of the Survey's investigations, which began to be published during the 1840's, most of the subdivisions used were pertinent only to the area for which they had been established, but Ebenezer Emmons did recognize a "Taconic System," which he claimed was applicable over a wide area, at the base of which were unfossiliferous slates and quartzites, overlain by the "New York System," which was, in turn, overlain by rocks considered "Old Red" (equivalent to the British Old Red Sandstone). In general, this was the terminology recognized by all, but they argued over the limits of the New York System and the advisability of delimiting the Taconic System. James Hall compared the faunas found in the "New York System" with those described from the Silurian and Devonian strata in England. Trilobites were later found in rocks beneath those bearing Silurian fossils and these were compared to those forms described by John Salter from Sedgwick's Cambrian. Thus, in North America, as in Europe, a distinctive fauna of trilobites was described from strata beneath those bearing Silurian faunas.

During the middle years of the nineteenth century, Charles Lapworth, a Scottish schoolmaster who was later Professor of Geology at Birmingham University, made many excursions to investigate the southern uplands of Scotland. He walked over the rolling hills, studying the forbidding graywackes, cherts, and shales that were piled and twisted in seeming confusion. Lapworth worked out the stratigraphic sequence in this area and noted that the rocks bore some fossils called graptolites, which had been little noticed before. By working out the order of the layers and the ranges of the several graptolite species through the layers, Lapworth was able to recognize distinctive assemblages of graptolites, and he called the rock layers bearing them, zones. Further work in the Lake District and in North and South Wales permitted him to piece together a sequence of several such graptolite zones within the slaty and shaly rocks that had previously yielded few fossils. He determined that some of these zones were in the Silurian System, and that some designated rocks that were older than those of the Silurian.

While Lapworth was building up his knowledge of the graptolite-bearing rocks, the Cambrian faunas were being described in many places. By the year 1879, the chief protagonists in the Cambrian-Silurian battle had died. Lapworth took the opportunity to point out, in an article, *On the Tripartite Classification of the Lower Palaeozoic Rocks,* published in the Geological Magazine for January, 1879, that the rocks included by Sedgwick and Murchison as Cambrian and Silurian bore (1879, p. 3–4):

> . . . *three distinct faunas,* as broadly marked in their characteristic features as any of those typical of the accepted systems of a later age. The necessity for a tripartite grouping of the Lower Palaeozoic Rocks and Fossils, in partial accordance with this fact, has been very generally acknowledged for the last thirty years.

Lapworth went on to show that within Murchison's original Silurian two faunas really existed, one typifying the Lower Silurian, and the other, the Upper Silurian. The third fauna was that collected from the lower part of Sedgwick's Cambrian rocks.

The Upper Silurian rocks and their fauna typified one period, the Silurian, maintained Lapworth. The lower parts of Sedgwick's Cambrian rocks had a distinctive fauna worthy of systemic recognition, and thus the Cambrian was a valid period. This consideration left the Lower Silurian rocks and their fauna without a name of systemic rank. To these rocks Lapworth gave the name *Ordovician System,* after the Ordovices, the last of the old British tribes to yield to the Roman Legions. The Ordovices had inhabited the area around the town of Bala in North Wales, and this was the district in which Sedgwick had first worked on the rocks that he termed Upper Cambrian and that Murchison considered to be Lower Silurian. This is the type area for the Ordovician.

Lapworth (1879, p. 14) designated the three systems in the following table:

> Silurian System: Strata comprehended between the base of the *Old Red Sandstone and that of the Lower Llandovery.*

> Ordovician System: Strata included between the base of the *Lower Llandovery* formation and that of the *Lower Arenig*.
>
> Cambrian System: Strata included between the base of the *Lower Arenig* formation and that of the *Harlech Grits*.

In this table, Lapworth definitively set forth the boundaries of each system. The faunas typifying each were already well known and widely recognized. The Ordovician Period was thus founded on a characteristic faunal aggregate.

Lapworth's suggested division slowly gained favor. The British Geological Survey, then under Archibald Geikie's directorship, did not immediately recognize the Ordovician, but when J. J. H. Teall succeeded him in 1900, the usage Lapworth suggested began to be followed. The United States Geological Survey officially adopted it in 1903, and the terms were included by Thomas Crowder Chamberlain and Rolin D. Salisbury in their widely used textbook, which was published in 1906. Since that time, Lapworth's usage has been followed by most of the geologists in America and England. On the Continent, however, Silurian has been widely discussed as a Period with a Lower and an Upper Silurian as divisions. The Lower Silurian in such usage is the same as Lapworth's Ordovician—not the Lower Silurian as defined by Murchison either in 1835 or after he revised his original ideas. The Upper Silurian in such usage has also been called the Gothlandian. Gradually, however, more and more geologists are using the Lapworth terminology, and the Ordovician is well established as a period.

The Pennsylvanian and the Mississippian

A basic twofold subdivision of the Carboniferous Period has been accepted by most European workers. North American stratigraphers have, however, divided what was formerly designated as the Carboniferous System into two systems: Coal-bearing beds were established as the type of the younger of the two, the Pennsylvanian System, and limestone deposits, the older, the Mississippian System.

Coal Measures were recognized early in the exploration of the United States, and they were first studied intensively in Pennsylvania. Active geologic interest in the United States was fostered by state geological surveys. The first was established in New York in 1836. A Pennsylvania State Geological Survey was formed in the same year. Unhappily, it was forced to disband after six years because of financial stress. Its principal geologist, H. D. Rogers, continued privately to work on his reports, and presented them to the state legislature, which finally published them in 1858 after some bickering about costs and worthiness and delays. Rogers' study of the Pennsylvania Coal Measures established them as the best known in the country. In the course of his work, he was able to point out five subdivisions of the Coal Measures that were widely applicable within the state. Four of these five remain today as a set of subdivisions of the Pennsylvanian age rocks in Pennsylvania.

While the Pennsylvania Survey was getting started, geologic inquiry was spreading into the northern part of the Mississippi River Valley. There, workers recognized a sequence of fossiliferous limestones beneath the Coal Measures. In 1839, David Dale Owen applied the term "Upper Carboniferous" to the Coal Measures and the term "Subcarboniferous" to the limestones beneath them, defining them as subunits of the Carboniferous age rocks. In succeeding years, geologic investigation continued in this area. In 1870 Alexander Winchell used the term "Mississippian" for the calcareous rocks that had been demonstrated to underlie the coal in the northern part of the Mississippi River Valley. In Winchell's use, the term "Mississippian" meant the calcareous facies or aspect of the rocks under the coal, to emphasize the contrast with the sandstones and shales that underlay the coal farther to the east.

By about 1890, stratigraphic nomenclature in the United States was extensive and confusing. To reduce the confusion, the United States Geological Survey decided to bring all that was known about the stratigraphy of the country up to date in a set of publications called *Correlation Papers*. Henry Shaler Williams proposed in his discussion of the Carboniferous correlations in 1891, the recognition of the Mississippian as a series (of rocks) which were correlative with the lower part of the Carboniferous in Europe. He indicated that the Mississippian was a series of dominantly calcareous rocks beneath the Coal Measures and above the Devonian age rocks and that an exposure in the upper part of the Mississippi River Valley provided a type area. In the same paper he also proposed that the name "Pennsylvanian Series" be used for the coal beds that were best known in the state of Pennsylvania. These two terms were thought to be subdivisions of a Carboniferous System. The Mississippian and Pennsylvanian were elevated to systems in Chamberlain and Salisbury's *Textbook of Geology*. These authors believed that the difference in lithologic aspect between the Pennsylvanian and Mississippian rocks was sufficiently great to warrant their being recognized as distinct systems (and periods). They also argued that an unconformity separated the two.

The Chamberlain and Salisbury text was published at a time when others available in North America were outdated, and it was widely used: Most American geologists who were educated during the period from 1904 to 1930 learned their basic geology from it. The Chamberlain and Salisbury use of Mississippian and Pennsylvanian was widely circulated and taught, and thus came to be widely accepted in North America. The majority of the state geological surveys were using the Chamberlain and Salisbury suggestion by the 1930's, but not until

May 14, 1953, did the United States Geological Survey officially recognize the Mississippian and Pennsylvanian as distinct systems (and periods). The two are almost unanimously accepted as periods in the United States, but, despite the arguments of American stratigraphers, geologists in the rest of the world still use the single period, the Carboniferous.

CHAPTER SEVEN # Lyell's Percentages

Smith's principle of faunal succession led gradually to recognition of large aggregates of fossils as the bases upon which the periods were founded. As the periods were being established, the need was recognized for divisions of lesser magnitude. Charles Lyell proposed a method for the analysis of relatively large fossil aggregates which led him to establish the epochs of the Tertiary and Quaternary.

Lyell made comparisons of the numbers of still-living species of mollusks and extinct species in successively higher stratigraphic positions within the Tertiary and Quaternary. He discovered that the stratigraphically lower beds were characterized by relatively few still-living species and the higher by relatively many. This discovery led him to formulate the principle that the percentages of still-living species in large aggregates of fossils collected from Tertiary rocks could be used for subdividing the Tertiary. It must be emphasized that Lyell studied relatively large fossil aggregates collected from many stratal units while formulating his principle. Lyell did not concern himself with precise stratigraphic ranges of individual species: He dealt with bulk faunas from many stratal units. Lyell was also not concerned with the boundaries of the time divisions he proposed. His time intervals were not precisely delimited.

Of all the recognized subdivisions of the periods, perhaps none have attracted more attention from geologists than Lyell's epochs and series of the Tertiary. Lyell proposed his subdivision of the Tertiary before many of the other periods had been established. Because of this and because of the wide attention they have received, Lyell's Tertiary subunits, which he originally termed "periods," but which today are known as "epochs," have been considered to be of nearly the same importance as the periods by some students of earth history.

Charles Lyell stands in the front rank of the founding fathers of geology for it was he who amplified Hutton's presentation of the principle of uniformity of nature's processes, and he, who, by a lifetime of study of natural phenomena, made that principle a cornerstone of the field. During his early geologic inquiries, he traveled to many of the Tertiary deposits in Europe. After these travels, he formulated his principle for subdividing the Tertiary and for correlating deposits within it.

While at Oxford, Lyell took one of Professor Buckland's courses in geology and spent a summer touring the Continent looking at glaciers and collecting fossils. After leaving Oxford, he moved to London and began to study law at Lincoln's Inn. At the same time, in the fall of 1819, he became a Fellow of the Geological Society of London and also of the Linnaean Society. At meetings of these societies, he became fast friends with some noted scientists, and was particularly stimulated by those of his friends who were interested in geology. As his eyesight deteriorated, he discontinued the study of law, and traveled on the Continent and in England, making notes about rocks and fossils. He was elected Secretary of the Geological Society in 1823, and in the same year he read his first paper before that group. The subject was the geologic aspect of some rivers near his ancestral home in

Forfarshire, Scotland. During that year and the next, Lyell went on geologic excursions to the south of England and, in the company of Professor Buckland, into Scotland. He also went to France and met geologists there, including Brongniart, Cuvier, and von Humboldt. In 1825, at his father's insistence, he started to study law again, but continued his geologic investigations and began the manuscript that was to become his famous text, *Principles of Geology*, which stands today as one of the monuments in the literature of geology.

Late in 1827 Murchison asked Lyell to accompany him and his wife through France and northern Italy. Lyell accepted the invitation, and during most of 1828, he and the Murchisons toured the Continent, studying geologic phenomena and meeting geologists in the areas they visited. They traveled by carriage across the country, and used the modus operandi they had developed on earlier journeys: Mrs. Murchison busily sketched and hunted fossils while the men scurried to the hill tops and made arduous excursions on foot. They worked long and hard studying rocks and fossils. Late in the fall of 1828, Lyell left the Murchisons and turned across northern Italy and then went south to the tip of the country and on to Sicily. He stopped to visit geologists and fossil collectors, noting the resemblances these collections had to some of those he had seen in the Bordeaux country and other areas in France. He collected fossils to study later, examined the rocks, and discussed local geology with his hosts. Then Lyell turned north, went back across the length of Italy and on to Paris, arriving back in England by the end of February, 1829.

In the course of his travels in Italy and France, Lyell realized that the Tertiary strata could be subdivided. He noted in the introduction to the third volume of his *Principles of Geology* (1833, p. xii–xiii) that by January, 1829, he

> . . . had fully decided on attempting to establish four subdivisions of the great tertiary epoch, the same which are fully illustrated in the present work. I considered the basin of Paris and London to be the type of the first division; the beds of the Superga, of the second; the Subapennine strata of northern Italy, of the third; and Ischia and Val di Noto, of the fourth.

Lyell had seen that the beds were fossiliferous and had examined numerous collections of fossils from these strata. With the thoughts of subdividing the Tertiary in mind, Lyell arrived in Paris in February, 1829 and told Jules Desnoyers his intention of classifying the (1833, p. xiii) "different tertiary formations in chronological order, by reference to the comparative proportion of living species of shells found fossil in each." Desnoyers told Lyell that the conchologist Gerard Deshayes had much the same idea, which he based upon his studies of fossil and recent mollusk shells in museum collections. Lyell hurried to Deshayes' office with considerable excitement. Deshayes maintained that he could recognize three subdivisions, not four as Lyell held that he could. Lyell asked Deshayes to examine the Tertiary fossils he had collected in Italy, to prepare some tables showing the species of mollusks common to two or more Teritiary units, and to indicate those species known both as fossils in Tertiary strata and as living organisms. After Lyell returned to England, Deshayes started to work on the tables, and after two years' work, sent them to Lyell. Deshayes had examined more than 40,000 specimens representing nearly 8,000 species in making up the tables. He assigned each specimen to one of his three units. As Lyell studied the tables, he could delineate four subdivisions, but was willing to lump two into one and did so, using the word "Pliocene" in both of their names, when the subdivision of the Tertiary was published in the 1833 volume. Lyell noted that a great many shells had been examined, and that they had been studied only by the investigator who made up the tables. Because of this, differing identifications—which might have arisen if more than one person had studied the shells—had been eliminated. Further, Lyell documented (1833, Figs. 3, 4) the stratigraphic relationships of the large groups of strata from which the collections were made.

Under the heading "Subdivisions of the Tertiary Epoch," Lyell mentioned a Recent era as the period during which the earth had been inhabited by man. He noted that some rock layers contained evidences of man, and that these could with certainty be included as Recent formations. Although other strata might also be Recent, without traces of man a definite identification could not be made. Later in the book, Lyell described the Recent more fully—as a unit distinct from and younger than the Tertiary.

Lyell's four subdivisions are, from youngest to oldest:

Newer Pliocene period
Older Pliocene period
Miocene period
Eocene period

Lyell indicated that ninety percent of the species known as fossils in the Newer Pliocene strata were also known as living organisms. He pointed to the deposits at Val di Noto, Sicily, as bearing a typical Newer Pliocene fauna.

In the Older Pliocene strata, Lyell noted that (1833, p. 54):

> . . . the proportion of recent species varies from upwards of a third to somewhat more than half of the entire number; but it must be recollected, that this relation to the recent epoch is only *one* of its zoological characters, and that certain *peculiar species* of testacea also distinguish its deposits from all other strata.

He indicated that the formations in Tuscany and the Subapennine hills in northern Italy were typical of the Older Pliocene.

It is interesting to note that about a half century earlier Arduino had studied these same deposits while delineating the Tertiary, and that more than a century and a half before they had been the subject of Steno's investigations that resulted in his inducing the principle of superposition. To some extent at least, the type area of the Older Pliocene is also that of the whole Tertiary.

The Miocene Epoch was denoted by the fact that the proportion of living species to extinct ones was (1833, p. 54) "rather less that eighteen in one hundred." Lyell indicated that the

FIGURES 10 and 11. Two diagrams used by Lyell to illustrate the superpositional relationships of the Cretaceous Chalk (C), the Tertiary strata in the Paris and London Basins (d), the Tertiary strata in the Loire River Valley in the Touraine district, France, and at Superga Hill near Turin in the Piedmont district, Italy (e), and the Subappenine Beds in the Piedmont area, Italy (f). The beds in each of the areas bear mollusks that were studied by Deshayes. The superpositional relationships shown in the diagrams were used by Lyell to document the stratigraphic relationships of the molluscan faunas and thus to establish the superpositional relationships of his Eocene, Miocene, and Pliocene.

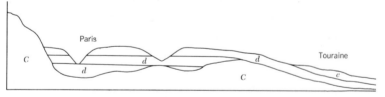

C: Chalk and other secondary formations (Cretaceous)
d: Tertiary formation of Paris basin (Eocene)
e: Superimposed marine Tertiary beds of the Loire (Miocene)

FIGURE 10. The stratigraphically highest beds of the Paris Basin Tertiary succession (d) are shown to be equivalent to strata that lie beneath Tertiary beds in the Loire River Valley, which bear a molluscan fauna with a greater percentage of still living species (e). [Redrawn from C. Lyell, *Principles of Geology*, vol. III, John Murray, 1833.]

Miocene formations were well developed in the Touraine basin, where the superpositional relationship to the underlying Eocene could be demonstrated, the area near Bordeaux in southern France, the basin around Vienna and adjoining parts of Hungary, and a part of northern Italy. The stratigraphic position of the Miocene beneath the Pliocene was worked out in northern Italy. Fossil aggregates from these several places were considered to demonstrate that the percentages were true generally. Apparently Lyell changed his mind to some extent concerning typical exposures of Miocene strata, for in 1829 he mentioned the deposits of the Superga hills near Turin in northwestern Italy as the type area, but in the 1833 volume he named instead the beds in the Touraine district in France known there as the "Faluns of the Loire," after the Loire River valley. This change has led modern stratigraphers to speculate about what should be taken as the type area for the Miocene. The Superga hills must be

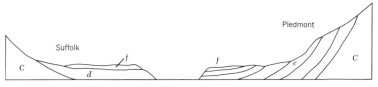

C: Chalk and older formations (Cretaceous)
d: London Clay (Older Tertiary: Eocene)
e: Tertiary strata of same age as beds of the Loire (Miocene)
f: Crag and Subappenine Beds (Pliocene)

FIGURE 11. The Subappenine Beds (f) are shown to overlie Tertiary strata (e) that bear a molluscan fauna similar to that of the Tertiary Beds in the Loire River Valley. Beds bearing the same fauna were recognized by Lyell at Superga Hill and near Bordeaux, France. The fauna recognized by Lyell in the Subappenine Beds was also found by him in the strata in England termed Crag. Lyell indicated that the fauna from the Subappenine Beds and the British Crag was younger than that of the beds in the Loire River Valley. [Redrawn from C. Lyell, *Principles of Geology*, vol. III, John Murray, 1833.]

recognized as the original type area, yet the "Faluns" have also been judged as the type area. The use of several areas by Lyell in establishing the Miocene proved valuable. The faunas analyzed are more diversified and their stratigraphic relationships are more completely documented than they could be from any single area. Lyell noted that the Eocene Epoch had a low proportion—approximately three and one-half percent—of living species within its strata. The Paris and London basins were discussed as the type areas for Eocene deposits and fossils.

Lyell did not consider that his subdivisions would necessarily be the only ones ever delimited within the Tertiary, for he stated (1833, p. 57–58):

> If intermediate formations shall hereafter be found between the Eocene and the Miocene, and between those of the last period and the Pliocene, we may still find an appropriate place for all, by forming subdivisions on the same principle as that which has determined us to separate the lower from the upper Pliocene groups. Thus, for example, we might have three divisions of the Eocene epoch,—the older, middle, and newer; and three similar subdivisions, both of the Miocene and Pliocene epochs. In that case, the formations of the middle period must be considered as the types from which the assemblage of organic remains in the groups immediately antecedent or subsequent will diverge.

Lyell erected a platform for future work on Tertiary rocks and faunas and pointed out that further analysis of the faunas might lead to other subdivisions of the same magnitude as those he had recognized. His units were based upon large faunal aggregates and the general superpositional relationships among the rocks bearing them.

Lyell interpreted the proportion of living species to extinct ones in a fauna as signifying a certain interval of time. A low proportion of living species in a fauna indicated time more distant from the present than a high proportion. Lyell pointed out that this method, or principle for interpreting faunal assemblages for time significance was only applicable to the Tertiary because only in Tertiary rocks are there a significant number of species identifiable with living species. He stated that it could be used widely within the Tertiary because, if fossil aggregates were collected in other lands, as in the Americas, for example, the percentages of living species in the faunas there could be determined. If only a small percentage were still living, then the deposit would belong in the oldest part of the Tertiary, the Eocene. If relatively high proportions of living species were found in the faunas, then the deposits from which they came would be in the upper part of the Tertiary or the Pliocene. Lyell noted that the proportion of extinct to living species in a fauna should be all that is needed to recognize the Tertiary subdivisions. It would not be necessary for the recognition of these subdivisions in the Americas that the species found there be the same as those found in the European areas.

Lyell himself emended his original terminology. In 1839, he restricted the word Pliocene to what he had called the Older Pliocene in 1833, and gave the term *Pleistocene* to what had been labeled Newer Pliocene. He thus established under four distinct names the four units he had recognized as subdivisions of the Tertiary.

A few years later, in 1846, Edward Forbes, in discussing the floral and faunal changes in England during the time of glaciation, stated that the glacial epoch was synonymous with the Pleistocene. Forbes's use of Pleistocene to mean the time of glaciation was widely circulated, and soon evidence of glaciation was accepted as the basis for the recognition of the Pleistocene. Lyell in 1873 changed his original definition of the Pleistocene and accepted Forbes's idea. Thus the basis for delineating the Pleistocene changed from faunal to climatic, and has been so defined by most geologists ever since. In such usage, the Pleistocene is not amenable to study in the same way as are subdivisions of time based upon faunas. Because of this, the Pleistocene deposits have been studied primarily by geologists other than stratigraphic paleontologists.

As Lyell had predicted might happen, more subdivisions of the Tertiary were recognized. In 1854 Heinrich Ernst von Beyrich delimited a subdivision between the Eocene and the Miocene. The upper part of the Eocene sequence recognized by Lyell in the Paris basin was not well developed and had yielded few fossils. In the northern part of Germany, in the area of the north German plain, and in Belgium, an extensive development of fossiliferous rocks was recognized which reflected an extensive marine transgression. These rocks were examined and subsequently shown to be contemporaneous with those in the upper part of the Eocene of the Paris basin. Von Beyrich discussed the nature of these deposits in north Germany and Belgium, and introduced the name *Oligocene* for them. Before the Oligocene Epoch was proposed as a distinct time interval, some geologists, of whom von Beyrich was one, had applied the designation "Lower Miocene" to these rocks, which were included by others in the Upper Eocene. This confusion was clarified when von Beyrich introduced the name Oligocene for a distinct time interval within the Tertiary. He stated that the Oligocene strata were developed differentially over northern Germany and Belgium, with few strata in some places and many more in others.

The lower part of Lyell's Eocene was also split off as a separate subdivision of the Tertiary. This was done in 1874 by W. P. Schimper in the course of his discussion of Tertiary floras. Schimper (1874, p. 680) said:

> I have grouped all the vegetation of the Tertiary into five distinct floras, which does not mean that I consider the five floras as independent of one another. All these floras are interrelated in time as our local floras are in space. But in spite of the evident

> continuity in the evolution of the organic kingdom through the geologic ages there can nevertheless be distinguished in this continuous and progressive movement a constant change in the grouping and relative development of types, a change which enables us to identify for each epoch, and even for each geologic period, a group of forms constituting what we call the organic character of the epoch or of the period.
>
> The group of plants or the flora of the period now under discussion, although allied closely with the Heersian flora, which is the continuation of the Cretaceous flora, and still more directly with the Eocene, has, nevertheless, an aspect peculiar to itself, which distinguishes it at a glance.

With these words, Schimper founded the Paleocene Epoch as an interpretive unit—one based upon a floral aggregate recognized to be unique in time. Schimper did note that the Paleocene floral aggregate was represented in two local floras from relatively restricted localities but that its position in the evolution of floras generally was intermediate between the floras characteristic of the Eocene and of the Cretaceous.

Schimper included the sands of Bracheux, the lignites and sands of Soissons, and the travertines of Sezanne in his Paleocene type. This group of rocks crops out a short distance northeast of Paris.

Although the Paleocene was founded upon a floral aggregate whereas the other Tertiary Epochs were recognized by analysis of marine invertebrates, some marine invertebrate fossils and some mammalian remains have been found in the type deposits of the Paleocene.

Lyell's subdivisions of the Tertiary stand historically among the first units of the geologic time scale to be based upon interpretation of faunas analyzed in stratigraphic order. The Lyellian percentage method of analyzing molluscan faunas is the guideline by which the interpretations were made.

Lyell did not establish boundaries for his units. Because of this, geologists, in attempting to subdivide and refine them, debate over their precise limits. For typological purposes and in the sense of Lyell's original meaning, such debates have little significance, for Lyell did not stipulate any means of recognizing the boundaries of his units nor did he precisely pinpoint all the deposits to be included in each type area. Boundaries between Lyellian units must be considered to be anything but sharp lines; they are broad and vaguely defined intervals.

Today, faunal comparisons involving a consideration of the stage of evolutionary development of certain lineages of organisms may be used to establish a position within the Tertiary, although this is commonly done within the context of a correlation using Lyell's principle, however unwittingly. To date, attempts to invalidate Lyell's principle have proven, upon careful study, to be erroneous. As study of mammalian distribution continues, several close correlations between widely separated areas may be possible by comparison of remains of land mammals. To be useable, the remains of land mammals should be found in beds also bearing plants or marine invertebrates. Once a correlation between land mammals and other organisms has been established, then, if the mammals lived and were preserved in many areas, time correlations between widely separated areas can be established. Of note is the fact that Lyell only considered the marine mollusks in establishing the Tertiary epochs. For typological purposes, other marine animals must be sought and related to the molluscan faunas, as must mammalian fossils, before either can be used for refined correlations of Tertiary rocks.

CHAPTER EIGHT **An Overview of the Earth's Fossil Record**

Historically, the Primitive, Transition, Secondary, and Tertiary were large descriptive rock units once thought to be broadly applicable in most of western Europe. They were in use when the interpretive units that today are termed periods were being established. As the interpretive nature of the periods came to be realized, the need for interpretive units of greater magnitude was recognized, and the eras were delineated. As the eras are based upon interpretation of the fossil record, they are interpretive units.

As geologists realized that periods based upon analysis of fossil aggregates were time intervals applicable only to that time span during which organismal remains were commonly preserved as fossils in the rocks of the earth's crust, they recognized that there is beneath the conspicuously fossiliferous rocks a vast portion that bears few fossils. This portion includes rocks of all kinds, and as geologic structures of the rocks in it were mapped, its history was realized to be very long. This portion of the earth's crust has been labeled "Precambrian" by many geologists. As its fossil remains are so sparse, time units based upon fossil aggregates cannot be delineated in it, with a possible rare exception here or there. The Precambrian is considered to be one era by most geologists concerned with fossiliferous rocks of the earth's crust; it will be so treated here. The Precambrian rocks have been divided into descriptive rock sequences in many areas. The problem of how to relate these local sequences remains essentially unsolved. Radioactive decay methods for determining past time provide the most hope for ascertaining the relationships between these sequences; such methods have only begun to be widely used.

The Eras

Sedgwick proposed the first of what we now know as the eras when, in 1838 in a talk before the Geological Society of London, he used "Palaeozoic Series" as the name for what is now known as the Paleozoic Era. Sedgwick attempted in that talk to show that two groups of stratified rocks were present beneath the Old Red Sandstone in England. He called the stratigraphically higher group the Paleozoic Series, and included within it his Cambrian Period and Murchison's Silurian Period. The rocks in this group were not as highly contorted or metamorphosed as the rocks beneath them.

In 1840, John Phillips, writing in the *Penny Cyclopaedia,* first used the terms "Mesozoic Era" and "Kainozoic [Cenozoic] Era" in the following manner (1840, p. 153–154):

> Supposing, as we think likely, that general terms for stratified rocks, thus formed upon a consideration of their organic contents, which appear to follow a great law of succession, will be preferred to others based on a view of their mineral qualities, which are certainly subject to repetition, there will be no other difficulty in their construction or application than what may be overcome by the progress of investigation. As many systems or combinations of organic forms as are clearly traceable in the stratified

crust of the globe, so many corresponding terms (as Palaeozoic, Mesozoic, Kainozoic, etc.), may be made, nor will these necessarily require change upon every new discovery.

Phillips included the Devonian System and Old Red Sandstone as well as the Silurian and Cambrian Systems in the Palaeozoic in this discussion of it.

The following year, Phillips' study of Devonian fossils from Cornwall, Devon, and west Somerset was published by the Geological Survey. In this work, he again referred to the Palaeozoic and the Mesozoic and, changing the "K" to a "C," to the Cainozoic. Phillips added to the Palaeozoic the Carboniferous rocks and the Magnesian Limestone (which Murchison later indicated was Permian). The Mesozoic was stated to include the Cretaceous, the Oolite (Jurassic), and the New Red Formation (the lower part of which was later demonstrated to be Permian, and the upper part, to be Triassic). The Cainozoic encompassed the Eocene, Miocene, and Pliocene "Tertiaries." Phillips probably used these latter terms in Lyell's original meaning, with the Cainozoic spanning the Tertiary to the Recent, and possibly even including the Recent. The words Palaeozoic, Mesozoic, and Cainozoic were based upon the Greek verb "to live" combined with the Greek word for "ancient," for "medial" or "middle," and for "recent," respectively.

Phillips amplified his original discussion of the basis for the three eras in these words (1841, p. 159–160):

> If we ask, in the same spirit, whether any dependence is certainly proved between the *antiquity of the strata and the forms of plants and animal remains* which they contain, we receive a satisfactory answer. There is proved to be a real and constant dependence of this nature, such that in every large region yet studied, where fossils occur in strata of very unequal geological age, there are whole groups of organic forms which occur only

in the oldest strata, others which prevail only in the middle, and some which are confined to the upper deposits.

If instead of classifying the strata by mineral or chemical analogies, we resolve to employ the characters furnished by successive combinations of organic life which have appeared and vanished on the land and in the sea, we shall obtain an arrangement of remarkable simplicity, more precise in application, and yet less disagreeably harsh in definition, than that which has been so long followed. We shall have three great systems of organic life, characterizable and recognizable by the prevalence of particular species, genera, families, and even orders and classes of animals and plants, but yet exhibiting, clearly and unequivocally, those *transitions* from one system of life to another, which ought to occur in every natural sequence of affinities, dependent on and coincident with a continuous succession of physical changes, which affected the atmosphere, the land, and the sea.

Phillips definitively made the eras interpretive divisions. Each was based upon a large group of fossils. The terms suggested by Phillips for designating the eras gradually found acceptance, which today is almost universal. The eras are convenient generalizations useful for distinguishing large segments of prehistoric time from one another.

The Precambrian

The rocks lying beneath the Cambrian System are commonly labeled "Precambrian" and promptly forgotten by most paleontologists and stratigraphers. Precambrian rocks have, however, attracted the attention of many eminent geologists, and the amount of time represented by the Precambrian is probably nearly seven times as great as the time that has elapsed since the beginning of the Cambrian. Precambrian rocks form the cores of continental areas and are widely distributed in many

- BELTS OF FOLDED ROCKS (POST-PRECAMBRIAN)
- FLAT ROCKS
- PRECAMBRIAN ROCKS

FIGURE 12. Structural map showing areas of exposed Precambrian rocks and areas that have been relatively stable since the Precambrian (flat rocks), and former mobile belts (belts of folded rocks). [As adapted from J. H. F. Umbgrove, *The Pulse of the Earth*, Martinus Nijhof, The Hague 1946, in Bernhard Kummel, *History of the Earth*, W. H. Freeman and Company, San Francisco, © 1961.]

parts of the world (Fig. 12). They are not amenable to the same method of subdivision as the rocks above them because they bear so few fossils. In certain areas some subdivisions based upon fossils have been worked out, as in Siberia, where fossil algae have been used. In most places, only physical (inorganic) criteria are available for delimiting subdivisions. Many local subdivisions have been established on a physical basis.

The most widely traced subdivisions of the Precambrian have often been termed "series," in the sense that they are series of formations. The Belt Series in the northwestern United States and adjoining Canada, the Grand Canyon Series in Arizona, and the Keewatin Series in Canada are examples of such series.

In addition to the series, units termed eras have also been recognized within the Precambrian. Wilmarth (1925, p. 14–23) has discussed most of the era terms suggested for use with Precambrian rocks and has shown how diverse opinions have been on dividing Precambrian time and rocks. Some of the Precambrian era terms have been used with different meaning by different authors. The Proterozoic, for example, was originally introduced as a group of series (series of formations), but it was later used to mean all of Precambrian time. Perhaps the simplest view of Precambrian time is to consider it one great era during which life originated and diversified although the record of those developments left in the rocks is slight.

CHAPTER NINE Units
with Boundaries

While the periods and Tertiary Epochs were being established on faunal aggregates collected from large groups of strata, many naturalists were studying local rock sequences and collecting fossils from them. Such local investigations enabled Lyell to grasp quickly the broad generalizations concerning subdivision of the Tertiary and, similarly, aided Murchison in constructing a map of European Russia after only two field seasons. Murchison also owed much to British naturalists whose studies increased his knowledge of the stratal and faunal succession of the Silurian in Wales.

In the course of such studies, formation names proliferated almost as rapidly as did fossil collections. The problem confronting stratigraphers in the mid-nineteenth century was how to synthesize the many local fossil collections and detailed stratigraphic sequences to form a broader picture of the earth's history. The local details were isolate pieces of a jigsaw puzzle lying in wait for some interested naturalist to put them together. Each piece could be related to one of the periods, but understanding the history of the earth during a period demanded smaller time units through the use of which the pieces could be related.

Smith's principle of faunal succession led only to the establishment of vaguely defined time units. Lyell's principle was applicable only to the Tertiary. It, too, permitted recognition of only vaguely defined time intervals. There existed no method for delimiting contiguous time units. A closer look at the evolution of organisms as seen in the rocks of the earth's crust was needed.

D'Orbigny's Stages

The energetic French paleontologist Alcide d'Orbigny attempted to provide an answer to the question of how to work out units shorter than the periods. While preparing a comprehensive survey of French paleontology and stratigraphy, he became dismayed at the vast number of local formations and fossiliferous sequences that had been studied, but had not been related to one another. In his work on Jurassic strata, *Terrains Jurassiques*, d'Orbigny noted (1842, p. 9) that:

> Geologists, in their classifications permit themselves to be influenced by mineralogic composition of the beds, whereas I take for my point of study, . . . the annihilation of an assemblage of organisms and replacement by another. I proceed solely on the identity of faunal composition. . . .

D'Orbigny recognized that similarity of faunal assemblage was the key to correlation of local rock units and their faunas. Faunas could be equated with similar faunas in other areas. In this, he followed in the footsteps of William Smith. D'Orbigny noted that certain groups of strata were characterized by fossil assemblages, or aggregates, peculiar to them. Such a group might include many formations in many different areas, or might include only a part of a formation in one place and all of it in another. What unified the rocks was the fossil assemblage. Consideration of lithologic aspects was brushed aside. D'Orbigny designated a group of strata bearing the same fossil assemblage, a *stage*. Further, he named most of the stages after geographic localities at which rocks bearing many of the fossils characteristic of each stage were found. In this regard, d'Orbigny stated that (1842, p. 604):

> One will see by the nomenclature adopted in naming the stages that I have tried . . . to take names drawn from the places where the stage is best developed, to put an end to confused nomenclature based on local mineralogic composition, which is so variable from place to place, and on fossils which may be dominant at one place but which may be lacking in others.

In summarizing his ideas on stages (at the end of his first volume on the Jurassic rocks in France) d'Orbigny noted that many subdivisions of Jurassic rocks had been proposed based on the lithologic aspects of strata or on the predominance of one fossil or another in some local sequence. Both of these kinds of subdivisions were misleading, hence he had turned to another solution of the subdivision problem. He had traveled widely in France and observed the succession of Jurassic faunas in many parts of the country. Further, study by other geologists of collections from England, Russia, Central America, and India had, he maintained, supported his observations concerning the succession of faunas. D'Orbigny asserted that he had always noted the same sequence of faunal assemblages in the same superpositional relationships, regardless of the lithologic aspect of the beds in which they were found. Because he had found the succession of faunas so constant over all of France and had some evidence of the same succession from other parts of the world, d'Orbigny was convinced that the faunas he had seen were to be found the world around, and that the stages bearing them were divisions of the rocks of the earth's crust that "nature has delineated with bold strokes across the whole earth." D'Orbigny's own observations had been accurate, and the stages could be recognized in Jurassic rocks throughout France, but, as later study showed, the stages and their characteristic faunas were not spread across the entire earth.

D'Orbigny was influenced by the Cuvier catastrophic ideas of earth history. To d'Orbigny, each of the faunas typifying one of the stages was the result of a creation following a catastrophe. The fauna characterizing each stage was annihilated by a catastrophe. In the d'Orbigny concept, a stage was a thickness of rock bearing a characteristic assemblage of fossils. The fossils were distributed around the world, and the rocks bearing them could be thought of as being like an onion. The demarcation between concentric layers was the appearance of a new fauna. In short, d'Orbigny's hypothesis was a sort of integration of an aspect of neptunism with catastrophism.

Although d'Orbigny exaggerated the geographic extent of his stages, within the area of his observational experience with Jurassic rocks and their faunas (that is, within France), the stages were valid. D'Orbigny's catastrophic ideas, which were popular throughout France in his day, were not acceptable to those naturalists who had read Lyell's amplifications of Huttonian uniformitarianism. Many English naturalists of the time knew that certain fossil species ranged through many rock units, and were found in different associations in different parts of their stratigraphic range. Despite these flaws in his hypothesis, d'Orbigny's contribution to the refinement of the geologic time scale has proven of value when tempered with the principle of evolution through natural selection and the knowledge of the stratigraphic ranges of species that are not confined to just one faunal aggregate. D'Orbigny's stages found quick acceptance by some geologists, notably Jules Marcou, Jules Thurmann, and Pierre Desor, who were studying Jurassic rocks in France and parts of Switzerland. They added more stages to the ten that the strata bearing the diagnostic aggregates, or congregations, of species Oppel gave the name *zones*.

Oppel (1856–1858, p. 4, 13, 15) pointed out that commonly only a few species—those appearing for the first time and those last seen in the zone—were members of the congregation. Many species ranged through two or more zones. Oppel specifically mentioned that it was the overlapping stratigraphic ranges of species that gave rise to the congregation which typified a zone. Oppel's principle, then, is that the analysis of the stratigraphic ranges of species leads to recognition of congregations of species that are unique in time. The boundaries between congregations are chosen at the first appearance of one or more members of the congregation, because these appearances are unique events in time. The time span corresponding to a zone, a phase, is the time that elapsed between the appearance or appearances chosen as the basal boundary of one zone and the appearance or appearances chosen as the basal boundary of the next succeeding zone. Oppel's principle of analysing the overlapping ranges of species thus permits recognition of clear-cut time units. No previous principle put forward for the analysis of fossils led to such definite time intervals.

Oppel expanded his discussion by grouping zones into stages (*Zonengruppen*). A stage, he maintained, was a larger unit than a zone as it was characterized by a larger magnitude fossil aggregate. Oppel's "stage" was thus slightly different from d'Orbigny's, based, as it was, on analysis of the overlapping ranges of species, which d'Orbigny did not use. D'Orbigny considered that each group of species had been specially created and had existed only for the time duration of its stage. Oppel knew that the stratigraphic ranges of species were variable, and used this knowledge to recognize the distinctive faunal aggregates upon which he based his stages.

Oppel actually divided the Jurassic into zones, and his zone

d'Orbigny had described. Following their work, many more stage names were proposed by other workers of whom Carl Mayer-Eymar was perhaps the most prolific in establishing new stages by recognizing slightly different faunal aggregates from place to place. The naming of new stages became more a sport than a scientific endeavor as most of the stages put forth by all the workers who followed d'Orbigny have proven to be meaningless variations or divisions of d'Orbigny's original ones.

While a few workers, most of whom were French and Swiss, engaged in the game of naming new stages, other geologists found that they were unable to recognize d'Orbigny's divisions of the Jurassic in their parts of the world. In Germany, Friedrich Quenstedt, as he studied a sequence of Jurassic rocks in Württemberg very carefully and in minute detail, measuring thicknesses of strata exactly and noting the precise stratal position of each fossil, found that he could not use the d'Orbigny stages; his criticism of d'Orbigny's method centered around the acceptance, as the diagnostic faunal aggregate, of species from many strata in many localities, lumped together without enough regard for their precise stratigraphic ranges. He maintained that geologists should begin their attempts to subdivide the periods with a very detailed study of the stratigraphic ranges of species, that they should examine all strata, centimeter by centimeter, for fossils, and note precisely where each fossil was found in each stratigraphic section. Then, analysis of these details would lead, through comparison of the ranges of the species in many stratigraphic sections in many localities, to a fuller understanding of the succession of faunas. From such an understanding, Quenstedt asserted, the subdivision of the periods might proceed.

Quenstedt was not the only geologist of his time who was interested in determining the precise stratigraphic ranges of species through rock sequences although he championed the cause more vociferously than others. Painstaking studies of the sort he proclaimed were necessary had been and were being carried out in many places. Louis Hunter (1836) and W. C. Williamson (1834, 1836, 1838) examined Jurassic strata exposed along the Yorkshire coast of England, measuring the stratal succession with care and noting precisely the stratigraphic ranges of the several species of ammonites found there. Their studies are examples of the kind of precise stratigraphic work being carried out in the 1830's and 1840's in many parts of Europe.

Oppel's Principle

Although Quenstedt had declared that an overall picture of stratal and temporal relationships of fossiliferous rocks could be obtained by comparing stratigraphic and fossil range details from many stratigraphic sections from many different areas, he worked in only a relatively restricted area and did not derive any widely applicable principle for doing so. The method by which all fossiliferous strata could be divided into small-magnitude, clearly delimited units was put forth by one of Quenstedt's students at the University of Tübingen, Albert Oppel. Oppel studied Jurassic strata and the fossils found in them during the 1840's and 1850's. He followed in his teacher's footsteps by carefully studying local stratigraphic details and determining the stratigraphic ranges of many species as precisely as possible. He began his studies in the Schwabian district of the Jura Mountains. Then, after measuring many stratigraphic sections in that area, he traveled widely in Europe, examining collections of Jurassic fossils and stratal sequences. From this extensive background of data concerning the Jurassic deposits and fossils in Europe, Oppel (1856–1858) formulated the principle which permits establishment of the small-magnitude units of the prehistoric time scale and also paved the way toward establishing clearly delimited time units of any magnitude in that scale. His work gave real substance to the contributions paleontologists can make t comprehension of the history of the earth.

Oppel began by determining the stratigraphic ranges many species of fossils as he could find in the Jurassic str posed in the Württemberg and Schwabian areas in Germar then extended his studies to include the Jurassic strata in F Switzerland, and England. He was careful to note as pr as possible the stratigraphic ranges of all the fossils foun ticularly the ammonites, in every stratigraphic section st Oppel took care (1856–1858, p. 3) to "investigate the v distribution of each individual species at many different ignoring the mineralogic character of the beds." He then the stratigraphic range or vertical stratigraphic distribut every species found in every stratigraphic section. From a of these plots, he quickly realized that there were gro strata characterized by closely similar fossil aggregates. lieved these bodies of strata were time correlative. Then, studied the stratigraphic ranges of the several species c making a sort of composite series of stratigraphic secti that as many species as possible might be included in his ar Oppel saw that the stratigraphic ranges of some species short (being found with only one aggregate), others wer long, and most were of some intermediate length. Speci peared and disappeared in a random manner in the stratig sections he had plotted. The different ranges of species le pel to realize that the distinctive faunal aggregates which c terize certain groups of strata were the result of the overl stratigraphic ranges of species. Using certain appearan new species, Oppel discovered he could quite clearly delin boundaries between the distinctive aggregates, and coul distinguish a succession of distinctive aggregates, each of was bounded at its base by appearances of certain new s and at its top by appearances of certain other new speci boundaries at the base and top of the Jurassic System were also system boundaries. Oppel's principle thus permits use of zone boundaries—the first appearances of new species—as particular events in time that may be used to demarcate time intervals. Zone boundaries so recognized may coincide with stage, series, and system boundaries.

About a year after Oppel's work was completed, Darwin's contribution—evolution through natural selection—made clear why the vertical stratigraphic ranges of fossil species overlap. Subsequent to Darwin's contribution, at least some geologists have realized that it is evolution through natural selection that is the ultimate basis for their time scale based on fossils, and that as events in it are unique in time, they may be used to delimit time intervals. All subsequent refinements of Oppel's principle have involved comprehension and use of Darwin's contribution.

Some Modern Modifications of Oppel's Method

Although the basic principle put forward by Oppel has remained unchanged through the years since his formulation of it, some modifications of his method of studying and interpreting the data provided by the presence of fossils in rocks have arisen. These modifications have come about primarily as the studies of the stratigraphic ranges of fossils have included all parts of the geologic succession (not just the Jurassic) in many parts of the world. The discussion will center upon the basic unit recognized by Oppel, the zone. This unit is a time-stratigraphic unit, that is, it refers to all the rocks deposited during the time span of the corresponding phase. Stage, as used by Oppel (group of zones), is also a time-stratigraphic unit, that is, it refers to all the rocks deposited during the time span of the corresponding age. Zones and stages are most useful in determining precise relationships between events in earth history. Lithofacies maps and isopach diagrams, for example, have their greatest meaning when they are constructed for particular zones.

Perhaps the most important addition to Oppel's method of determining zones is the recognition that definite geographic limits exist within which a time-stratigraphic zone may be traced and beyond which it may not be recognized. The area in which a zone may be recognized is a biogeographic province. The organismal group examined for purposes of zonation must also be analysed to ascertain its biogeographic distribution pattern in order to establish biogeographic provinces. The overlapping vertical stratigraphic ranges considered in the steps toward delineating time-stratigraphic zones are thus the ranges within a biogeographic province and so rock sequences throughout the province must be sampled to work out accurately the vertical stratigraphic range of any species. Each province thus has a set of time-stratigraphic zones (and phases) that are unique to it and which must be correlated with those of other generally contemporaneous provinces.

An example of biogeographic provinces that have been established each with its own set of zones is to be found in the study of Jurassic ammonites by W. T. Dean, D. T. Donovan, and M. K. Howarth (1961). They recognized an Early Jurassic Northwest European Ammonite Province and a Mediterranean Ammonite Province. Study of the Early Jurassic ammonites revealed both the provinces and a set of zones based upon ammonite vertical stratigraphic ranges that were restricted in their applicability to each province. It is to be emphasized that the faunal analysis included both a consideration of the geographic extent of the diagnostic associations of each zone and the vertical stratigraphic range of each species.

A second modification in the delineation of time-stratigraphic zones made since Oppel's time is that in delimiting the species studied. Today, an attempt is made in working with fossil spe-

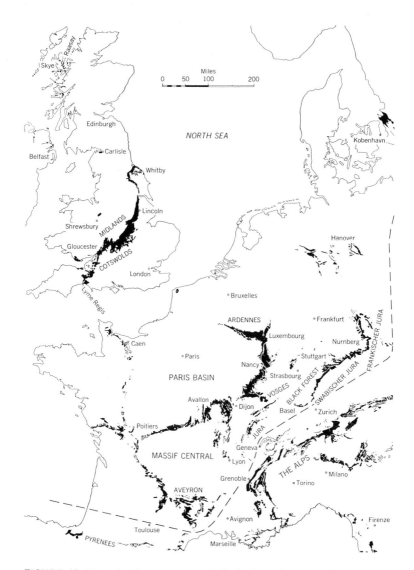

FIGURE 13. Map showing outcrops of Early Jurassic strata in northwest Europe. The dashed line marks the southern and eastern boundary of the Northwest European Liassic (Early Jurassic) Ammonite Province. The Mediterranean Liassic Ammonite Province lies east and south of the line. [Adapted from W. T. Dean, D. T. Donovan, and M. K. Howarth, 1961, The Liassic Ammonite Zones and Subzones of the Northwest European Province: British Museum (Natural History) *Geology Bulletin*, vol. 4, no. 10, pl. 75, 1961. By courtesy of the Trustees of the British Museum (Natural History).]

cies to place them in proper position in ancestor-descendant (phyletic) lineages. Principles of genetics must be invoked in working with fossil remains because even though the species recognized by paleontologists must be based upon morphological characteristics and degrees of morphological similarity, genetics indicates that morphological similarity among organisms living today is a reflection of similarity in gene composition and ability to interbreed successfully. Narrowly and unnaturally delineated species do not reveal organismal evolutionary development and thus obscure the evolutionary patterns that provide the primary basis used for the discernment of the passage of past time.

The invasion of species in lineages of the organismal group on which the fossil biogeographic province is based that are new to the province, from another, and the migration of species from one province to another, are now known to be just as significant for the purposes of time-stratigraphic zonation as the development of new species from existing phyletic lines and the extinction of old lineages within a province. Both the phyletic and the biogeographic appearances and extinctions may be used as events in time considered to be the bases for time-stratigraphic unit boundaries. The species new to a province through invasion of it from another may be short-lived, or they may give rise to established lineages. In either case, their appearance in a province is a distinct event in the history of the province and as such may be used for the purposes of recognizing time-stratigraphic units in the same manner as the appearance of new species from ancestor-descendant lineages in existence within the province. The distinction between the two kinds of appearances should be made because that related to invasion from another province may aid in correlating the time-stratigraphic units in one province with those in the province from which the organisms came.

The inclusion of fossils from deposits formed under many

different environments in analyses for time-stratigraphic zonal congregations is another important factor emphasized in the modern delineation of time-stratigraphic zones. This use of species collected from many different types of environments, permits all possible phyletic lineages of the organismal group to be analysed. Although the problem of recognizing time-stratigraphic units that will be generally applicable throughout a province and not to those deposits of just one particular kind of environment may not be so marked in work with planktonic as with benthonic organisms, planktonic organisms do exist in communities and some are closely related to benthonic communities. Indeed, many are but members of a community which includes both planktonic and benthonic members. Possible environmentally significant relationships between fossils of planktonic organisms and rocks of different lithologic aspect cannot be ignored.

Zone Names

According to Oppel's (1856–1858, p. 813) method for designating zones, one species in a zone is chosen, arbitrarily, to give its name to the zone. The species whose name is used may or may not be limited in its vertical stratigraphic range or in its biogeographic range to the zone. Further, it may or may not be well represented in the zone, although it commonly is one of the species that typifies the zone.

The characteristic feature of a zone is its congregation of species. No single species, not even the one providing a name for the unit, is an "index" or absolute guide to a zone delineated by Oppel's principle. The use of a time-stratigraphic zone name should be restricted to the province in which the zone was delineated unless the diagnostic congregation can be demonstrated in more than one province.

"Zone"—Other Uses

The word "zone" has been widely used in geology with many meanings other than "a time-stratigraphic unit." It has been descriptively applied many times—for example, to a body of rock with an observed, distinctive lithologic aspect, or a biologic aspect, such as an oyster bank. Unhappily for geologists the descriptive uses of the word "zone" outnumber the interpretive uses, and the many meanings can be confusing (Berry, 1966). Time-stratigraphic zones are interpretive units and can be recognized only after analysis of the distribution of fossil species collected from stratigraphic sequences.

Miocene Zones and Stages in California

Following Oppel's example, geologists working with the strata and faunas of other systems than the Jurassic have delimited zones and stages. Oppel's principle has been widely applied in Europe, and has been used, although to a lesser extent, in other parts of the world. It was used by W. Leupold and I. M. van der Vlerk (1931) and by J. H. F. Umbgrove (1938) in their syntheses of the Tertiary history in the former Dutch East Indies. They used zones, which were worked out using Oppel's principle, in the analysis of foraminiferal faunas. The studies by Leupold and van der Vlerk and by Umbgrove were particularly significant, for they further documented the validity of Oppel's principle.

Oppel's principle has had little application in the United States as yet. Indeed, whenever zones and stages have been delineated in the United States using his principle, considerable controversy has surrounded them because few geologists are aware of the implications and uses of these units (see discussion by Berry, 1966). One example may serve to describe this situation.

A sequence of zones and stages erected within the Miocene Series in California by R. M. Kleinpell in the early 1930's has been of considerable interest and importance to petroleum geologists. A similar study for the lower part of the California Tertiary succession was done by V. S. Mallory (1959). In both studies, foraminiferal evolutionary development was intensively analyzed, and stages and zones were based upon certain events within this evolutionary development. Kleinpell (1938) subdivided the Miocene into six stages, which are, from oldest to youngest, Zemorrian, Saucesian, Relizian, Luisian, Mohnian, and Delmontian. Zones were established within each stage. Care was taken to include foraminifers from rocks that had been deposited under different environmental conditions so that the representation of the faunal aggregates of the stages and zones would be as complete as possible. The recognition of the congregations upon which the zones and stages were based was carried out following Oppel's methods and the current modifications of them. These units have proven to be generally applicable to study of the west coast Miocene foraminifer-bearing deposits. Although subject to some modification (mainly the extension of certain species' ranges) as new information is acquired, they are, in essence, well tested. This succession of Miocene stages and zones in the California Tertiary was worked out at a time when most North American stratigraphers believed that such units were not applicable to North American rocks and faunas, with the exception of the physical divisions of the Pleistocene based upon glacial advance. Some geologists would not even consider stages or zones based upon Oppel's faunal-analysis principle as possibilities. When the description of Miocene stages and zones in California was being readied for publication, the publisher decided not to print it. This was in 1933, and a new code of American stratigraphic practice had just been published. The code was followed closely by all American stratigraphers. It stated that stages had no place in North American stratigraphy; zones were not even mentioned. Official geologic organizations, such as the U. S. Geological Survey and the Geological Society of America, were loath to publish material not encompassed by the new code. A geologic principle, demonstrated as valid in many studies made in western Europe and in other places, such as the East Indies, and shown to be applicable to North American faunas and rocks, could not be accepted because the code legislated against certain terms and, in essence, against scientific principle as well.

Five years passed before some petroleum geologists persuaded the American Association of Petroleum Geologists to publish the results of the study of California Miocene rocks and their faunas. The use of the stages and zones had paid off economically in the meantime in private practical applications. Kleinpell's study stands as one of the models of precise interpretive investigation of North American geology. Surprisingly, such studies are few in American geologic annals. American geologists have commonly held the results of their stratigraphic paleontologic investigations in straitjackets of observation and description—on occasion, of the purest abstraction—and have published little that may be considered substantiated interpretation.

Although it has not yet been recognized and used by all stratigraphic geologists, Oppel's principle provides a sound basis for analyzing the vertical stratigraphic ranges of fossil species in the rocks of the earth's crust. His principle, the analysis of overlapping stratigraphic ranges of fossil species and the selection of certain appearances of new species as unique events in time, can lead to establishment of clearly delimited time units.

CHAPTER TEN # Marker Points and the Correlation Web

To this point, we have traced the growth and development of a time scale based on organic evolution. It had its beginnings in purely descriptive rock groups, but, after fossils became acknowledged as valuable to recognizing the passage of time, the interpretive units were developed.

Faunal succession and Lyell's percentage analysis of the Tertiary fossils permitted only relatively long, vague time units to be recognized. Oppel's principle established some rather more clear-cut time units and paved the way to the realization that time units could be delimited precisely using fossils, as there were specific events in the evolution of organisms that could denote the beginnings of time intervals. The appearances of new species, the invasions of species from one faunal (or floral) province to another—these are events in organismal development that may be used as points in time, as, for example, a marker point or peg at the beginning of one time interval. Such a time interval lasted until certain other events occurred and left their record in the rocks, denoting the beginning of the next time interval.

Today it is realized that the selection of a boundary of a time interval should be made by choosing a definitive point in a stratigraphic succession in a particular place. Ideally, if fossils are used for such points, the points are chosen in places where the stratigraphic succession is continuous and the record of at least one evolving lineage is complete. The point picked as the beginning of a time interval is thus established in a type section. It is the task of geologists to spin a web of correlations with that point.

As he uses fossils to pick such points and to spin such a web the geologist must be wary of many pitfalls. To establish time intervals based upon the processes of organic evolution the geologist must: 1) collect fossils in stratigraphic succession; without this step, the meaning of the fossil sequence will be lost and evolutionary development may not be recognized; 2) determine, through comparative morphological analysis (and using as fully as possible pertinent principles of comparative anatomy, embryology, and genetics) and the knowledge of the superpositional order of the fossils, their ancestor-descendant relationships and their natural grouping into species (groups of organisms embodying common anatomical traits that are effectively isolated reproductively); these procedures demand conversance with biological principles; 3) define a boundary at the beginning of a unit based upon speciations or invasions of species from one area to another. The appearance of a new species among the trilobites (or several such appearances), for example, may be used to denote the beginning of the Ordovician and hence the Cambrian-Ordovician boundary. The appearance of the trilobite species is included in the Ordovician. That time unit can be concluded to have existed from that appearance to the time of appearance of certain other new species which denote the beginning of the Silurian. The new appearances in each case will be those at a particular point in a stratigraphic succession in a certain locality.

The geologist is aware that the speciations he uses for denoting time boundaries may be recognized in particular areas only because organisms are dependent upon their environment, and their distribution is controlled by their interaction with it. Several synchronous speciations are thus needed to correlate precisely with a marker point used to denote a time-unit boundary that is based on one lineage. In moving away from the marker point, a web of correlations is built.

The bathymetric gradient presents (as do the latitudinal and altitudinal gradients) a sequence of slightly differing environments. So, too, do the different bottom types (sandy, muddy, and so on) that are present along the gradient. Events in the evolution of organisms that lived on, and in each position along, the gradient and the bottom must be correlated. These events must in turn be equated with similar events in the evolution of the organisms that floated and swam as well as those that were land based. After several such events have been equated, some time correlations may be established for what is known as a faunal province. Units established in one faunal province must be correlated with those in others (if such exist). The faunal provinces are areas of natural distribution of particular types of organisms: they may be recognized both among land based and aquatic organisms. For example, Australia is faunally distinct from other areas in being inhabited by marsupials, none of which are known to live elsewhere. It constitutes a faunal province distinct from North America, which is inhabited predominantly by placental mammals. The paleontologist interested in the geologic history of Australia would establish a set of time units based upon marsupial evolution. These units would be applicable only in Australia. A set of units for North America based upon evolution of placental mammals has been established. They are applicable in North America for that time span during which placental mammals have been in existence. These sets of regional units must be correlated after they have been established.

The correlations are built up slowly, and through a set of tedious steps.

The various time units based upon evolution are not of the same duration (even those that share the same position in the hierarchy of units). Some estimate of the relative duration of an age or phase, for example, to others may be made through analysis of the relative amount of evolutionary development that went on during the time intervals under study.

It is also realized that the correlations with a marker point do not result in the establishment of a definite time "line" or time "plane" extending outward from the marker point at its type position. Instead, the correlations lead to recognition of an interval that is approximately synchronous with the marker point. Within the context of the time that has elapsed since the beginning of the Cambrian, however, which is about 600 million years, the time span encompassed by such an interval is very short.

The methods of establishing and using time units from study of fossils may not appear as credible as the methods for reckoning the passage of time from the radioactive decay of certain isotopes of elements in some minerals, but fossils are more readily available and the methods are, when correctly used, sound and reliable. They are the more precise and more practical in dating marine deposits, which comprise the majority of the rocks with which earth historians work. Fossils and minerals from which radioactive decay dates may be obtained found in the same rocks permit the time scale based upon events in evolution to be related to years. Using such information, estimates have been made (Callomon, 1963; Miller, 1965) that the shortest units in the time scale based upon organismal evolution are intervals of about 375,000 to 500,000 years. Although such an interval may seem relatively long in terms of human experience, when the time that has elapsed since the beginning of the Cambrian is considered, the same interval seems remarkably short. To describe the precision that may be achieved by using fossils, J. H. Callomon (1963, p. 2) pointed out in his study of Jurassic faunas that the shortest time units in the Jurassic give a degree of accuracy that corresponds "to being able to distinguish [at the present day], for example an event in the year 1007 B.C. from one in the year 1000 B.C., a problem not without difficulties even in historic times."

The correct application of the method of telling time by use of fossils involves principles of structural and stratigraphic geology and many principles of biology. Essential to the entire procedure, it must be emphasized, is the collecting of fossils in stratigraphic sequence.

The geologist interested in applying the record of evolution to reckoning time in the prehistoric past may begin his work by growing pure strains of one species in the laboratory and observing its changes from generation to generation, or he may examine groups of several different species in differing environmental situations to determine the possible kinds of response to environmental changes. Such experimental observations may then be applied to similar changes or interactions revealed by the fossil record. If not actively engaged in research on living organisms, the geologist must not only comprehend such research, but he must also use, in terms of principle, its results. He should, therefore, be aware of the search for the origin of life and for the precise chemical mechanics behind the transmittal of inheritable traits, and cognizant of the principles of genetics, as background for determining natural relationships of the fossils he finds. Using information from biology and biochemistry in conjunction with the stratigraphic superpositional relationships of fossils, the geologist may make as accurate as possible his interpretations of their ancestor-descendant relationships. Ecology and biogeography also assist the geologist, for they open to him possibilities for exploration into interrelationships among many different kinds of organisms. From these interrelationships, he may learn more about

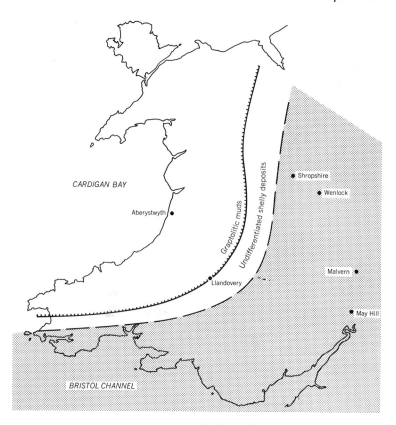

FIGURE 14. Early and Middle Llandovery paleogeography of Wales. (The Llandovery is the earliest Epoch of the Silurian Period.) The shaded area is land. The dashed line represents the approximate position of the shoreline at that time. The solid line with hachures represents the approximate position of the break in slope from continental shelf to slope. Note that the shelf is relatively narrow. This diagram does not show the general habitats of the brachiopod communities. [Adapted from A. M. Ziegler, Silurian Marine Communities and Their Environmental Significance, *Nature*, vol. 207, no. 4994, fig. 1, 1965.]

the restrictions environments place upon organisms. This information will be of help to him in interpreting his fossils. The record of changes in past environments sets a trap for the unwary geologist in which he may be caught as he analyzes his fossils for time significance, because rapid environmental changes are accompanied by rapid organismal changes. Such changes in the fossil record might be deemed significant enough to denote the beginning of one or more time units while all that should be interpreted from them is a rapid environmental shift during a particular interval of time. Faunal (and floral) provinces and the general patterns of organismal distribution must be sought and evaluated in using time units based upon evolution. Correlation between provinces may be one of the geologist's biggest headaches if the organisms that typify each province are not distributed in such a manner that there is an area in which members from each province overlapped in their geographic ranges. All of these essentially biological investigations invite the geologist's attention and offer challenging positions from which he may turn his steps toward using fossils to unravel the history of the earth.

In closing, perhaps one example in which time units based upon the evolution of organisms have been used to work out both an ecologic and a paleogeographic story during one part of one period (the Silurian) may reveal the contributions the close interpretation of fossils may make to understanding earth history. In a study of brachiopod communities from the rocks of Llandovery age (approximately Early Silurian) in Britain, A. M. Ziegler (1965) demonstrated that coeval stratal bands were characterized by different brachiopod communities (see Figs. 14–16) and were arrayed parallel to one another and to the shoreline that had existed at the time the rocks were forming. He concluded that the communities in each set of stratal bands that were approximately contemporaneous had been aligned along the bathy-

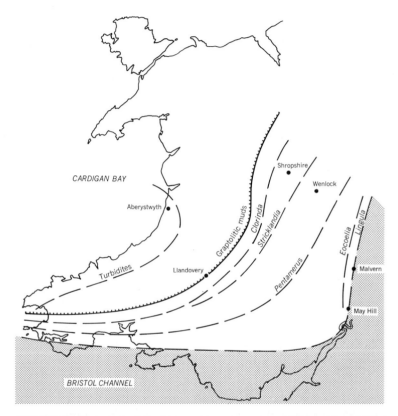

FIGURE 15. Early Late Llandovery paleogeography of Wales. The shaded area is land. The dashed line represents the approximate position of the shoreline at that time. The solid line with hachures represents the approximate position of the break in slope from continental shelf to slope. Note (by comparison with Fig. 14) that the shelf had greatly expanded since Middle Llandovery time. The brachiopod communities were well developed on the shelf. The position of the majority of the turbidite deposits is seaward from the foot of the slope. [Adapted from A. M. Ziegler, Silurian Marine Communities and Their Environmental Significance, *Nature*, vol. 207, no. 4994, fig. 2, 1965.]

FIGURE 16. Late Late Llandovery paleogeography of Wales. The solid line with hachures represents the approximate position of the break in slope from continental shelf to slope. Note that the position of this line remained constant throughout the Llandovery. The shelf had been even more greatly expanded from its Early Late Llandovery size (see Fig. 15), and all of Wales was under water. In the evolution of the pentameroid brachiopods, the Early Late Llandovery *Pentamerus* had given rise to *Pentameroides*, and in the evolution of the stricklandid brachiopods, *Costistricklandia* had developed from the Early Late Llandovery *Stricklandia*. The community types remained essentially the same even though evolution proceeded among some members of each. [Adapted from A. M. Ziegler, Silurian Marine Communities and Their Environmental Significance, *Nature*, vol. 207, no. 4994, fig. 3, 1965.]

metric gradient of the depositional area. He also pointed out that the replacement of the nearshore community at successively higher stratigraphic levels by communities indicating progressively deeper water showed that the shoreline had transgressed in an eastward direction during the Llandovery. The time intervals considered by Ziegler are founded upon the evolution of the brachiopods. Lineages of brachiopods in the communities have been studied and speciation events in these lineages used for the reckoning of time.

Ziegler's study is just one of the many in which a geologist has used fossils to help him decipher the history of the earth—a history that would be incomprehensible were there no way of measuring geologic time. As our understanding of organic evolution increases, the intervals in the scale of time units become shorter, and more of earth's secrets are revealed. The time scale based upon evolution is continually developing toward greater refinement: this book is but a progress report on its growth.

Literature Cited

Alberti, F. A. von, 1834, Beitrag zu einer Monographie des Bunter Sandsteins, Muschelkalks, und Keupers und die Verbindung dieser Gebilde zu einer Formation: Stuttgart und Tubingen, Verlag der J. G. Cotta'schen Buchhandlung.

Arduino, Giovanni, 1760, a letter to Sig. Cav. Antonio Valisnieri, in Nuova raccolta di opuscoli scientifici e filologici del padre abate Angiolo Calogierà: Venice, V. 6, p. 142–143.

Arkell, W. J., 1933, The Jurassic System in Great Britain: Oxford, Clarendon Press.

―――, 1956, Jurassic geology of the world: Edinburgh, Oliver and Boyd.

Barrande, Joachim, 1852–1911, Systeme Silurien du centre de la Boheme: Prague, Charles Bellman, 8 vols., 29 parts.

Berry, W. B. N., 1964, The Middle Ordovician of the Oslo region, Norway: Norsk. Geol. Tidsskr., v. 44, p. 61–170.

―――, 1966, Zones and zones—with exemplification from the Ordovician: Am. Assoc. Petroleum Geologists Bull., v. 50, p. 1487–1500.

Beyrich, H. E. von, 1854, Ueber die Stellung der hessischen Tertiarbildungen: K. Preuss. Akad. Wiss. Berlin Monatsber., November 1854, p. 664–666.

Brongniart, Alexandre, 1821, Notice sur des vegetaux fossiles traversant les couches du Terrain Houiller: Annales Mines, v. 6, p. 1–15.

Buch, C. L. von, 1839, Uber den Jura in Deutschland: Abh. Kg. Akad. Wiss. Berlin (dated 1837, published 1839).

Buckland, William, 1823, Reliquiae Diluvianae; or observations on the organic remains contained in caves, fissures, and diluvial gravel and on other geological phenomena attesting the action of an Universal Deluge: London, J. Murray.

Buffon, Georges Louis Leclerc, Comte de, 1781, Natural History general and particular by the Count de Buffon: William Smellie trans., v. 1, London, W. Strahan and T. Cadell; Edinburgh, W. Creech.

Callomon, J. H., 1963, The Jurassic Ammonite-faunas of East Greenland: Experientia, v. 19, p. 1–6.

Chamberlain, T. C., and Salisbury, R. D., 1906, Textbook of geology: New York, H. Holt.

Conybeare, W. D., and Phillips, William, 1822, Outlines of the geology of England and Wales: London, William Phillips, George Yard.

Cuvier, Georges, 1817, Essay on the theory of the earth (3rd ed.); Edinburgh, W. Blackwood.

Darwin, Charles, 1859, On the origin of species by means of natural selection, or the preservation of favoured races in the struggle for life: London, J. Murray.

Dean, W. T., Donovan, D. T., and Howarth, M. K., 1961, The Liassic Ammonite zones and subzones of the north-west European province: British Mus. Nat. Hist. Bull., Geology, v. 4, no. 10, p. 437–505.

Desnoyers, J. P. F. S., 1829, Observations sur un ensemble de dépots marins plus récens que les terrains tertiaires du bassin de la Seine, et constituant une formation geologique distincte; précédées d'un apercu de la non simultanéité des bassins tertiaires: Annales sci. nat., v. 16, p. 171–214, 402–491.

Forbes, Edward, 1846, On the connexion between the distribution of the existing fauna and flora of the British Isles, and the geological changes which have affected their area, especially during the epoch of the Northern Drift: Great Britain Geol. Survey Memoir, v. 1.

Füchsel, G. C., 1761, Historia terrae et maris ex historia Thuringiae per montium descriptionem: Akad. gemeinnutziger Wiss. zu Erfurt.

Giraud-Soulavie, Abbé, 1780, Histoire naturelle de la France méridionale: Paris, Nismes, v. 1.

Gould, S. J., 1965, Is uniformitarianism necessary?: Am. Jour. Sci., v. 263, p. 223–228.

Hill, R. T., 1887, The topography and geology of the Cross Timbers and surrounding regions in northern Texas: Am. Jour. Sci., 3rd ser., v. 33, p. 291–303.

Humboldt, Alexandre von, 1799, Ueber die unterirdischen Gasarten und die Mittel ihren Nachtheil zu vermindern: Braunschweig.

Hunter, Louis, 1836, Accompanying remarks to a section of the Upper Lias and Marlstone of Yorkshire, showing the limited vertical range of the species of Ammonites and other Testacea, with their value as geological tests: Geol. Soc. London Proc., v. 2, no. 46, p. 416–417.

Hutton, James, 1795, 1802, Theory of the earth, with proofs and illustrations: v. 1 and 2, Edinburgh, W. Creech (1795); v. 3, London, Cadell and Davies (1802).

Kitts, D. B., 1966, Geologic time: Jour. Geology, v. 74, p. 127–146.

Kleinpell, R. M., 1938, Miocene stratigraphy of California: Tulsa, American Association of Petroleum Geologists.

Kulp, J. L., 1961, Geologic time scale: Science, v. 133, no. 3459, p. 1105–1114.

Lapworth, Charles, 1879, On the tripartite classification of the Lower Paleozoic rocks: Geol. Mag., n. ser., v. 6, p. 1–15.

Lehmann, J. G., 1756, Versuch einer Geschichte von Flötz-Gebürgen: Berlin, Klüterschebuchhandlung.

Leupold, Wilhelm, and I. M. van der Vlerk, 1931, The Tertiary: in Feestbundel K. Martin, Leidsche geol. Mededeelingen, pt. 5, p. 611–648.

Lyell, Charles, 1830–1833, Principles of geology: London, J. Murray (v. 1, 1830; v. 2, 1832; v. 3, 1833).

———, 1839, Elements of Geology, French Translation: Paris, Pitois Levrault.

———, 1873, Antiquity of man (4th ed.): London, J. Murray.

Mallory, V. S., 1959, Lower Tertiary biostratigraphy of the California Coast Ranges: Tulsa, American Association of Petroleum Geologists.

Mantell, G. A., 1822, The fossils of the South Downs; or illustrations of the geology of Sussex: London, L. Relfe.

Miller, T. G., 1965, Time in stratigraphy: Paleontology, v. 8, p. 113–131.

Morlot, Andre, 1856, Notice sur le quaternaire en Suisse: Soc. vaudoise sci. nat. bull., v. 4, p. 41–45 (read March 15, 1854).

Murchison, R. I., 1835, On the Silurian System of rocks: London and Edinburgh Philosophical Magazine and Journal of Science, 3rd ser., v. 7, p. 46–52.

———, 1839, The Silurian System: London, J. Murray.

———, 1841, First sketch of some of the principal results of a second geological survey of Russia, in a letter to M. Fischer: Philos. Mag., v. 19, p. 419–422.

———, 1852, On the meaning of the term "Silurian System" as adopted by geologists in various countries during the last ten years: Geol. Soc. London Quart. Jour., v. 8, p. 173–184.

———, 1854, Siluria: London, J. Murray.

———, Edward de Verneuil, and A. von Keyserling, 1845, The geology of Russia in Europe and the Ural Mountains: London, J. Murray, 2 vols.

Omalius d'Halloy, J. G. J. d', 1822, Observations sur un essai de carte géologique de la France, des Pays-Bas et des contrées voisines: Annales Mines, v. 7, p. 353–376.

Oppel, Albert, 1856–1858, Die Juraformation Englands, Frankreichs und des südwestlichen Deutschlands: Württemb. Naturwiss. Verein Jahresh., v. xii–xiv (p. 1–438, 1856; 439–694, 1857; 695–857, 1858) Stuttgart.

Orbigny, A. D. d', 1842, Paléontologie Francaise, Terraines Jurassiques, Pt. 1, Cephalopodes: Paris, Masson.

———, 1849–1852, Cours elementaire de paléontologie et de géologie stratigraphiques: Paris, Masson, 2 vols. and atlas.

Owen, D. D., 1839, Report of a geological reconnoissance of the State of Indiana made in the year 1837 in conformity to an order of the legislature: Indianapolis, J. W. Osborn and J. S. Willets.

Phillips, John, 1840, Palaeozoic Series, in Long, G., ed., the penny cyclopaedia of the society for the diffusion of useful knowledge: London, Charles Knight, v. 17, p. 153–154.

———, 1841, Figures and descriptions of the Palaeozoic fossils of Cornwall, Devon and east Somerset: London, Longman, Brown, Green, and Longmans.

Playfair, John, 1802, Illustrations of the Huttonian theory of the earth: Edinburgh, W. Creech.

Quenstedt, F. A., 1856–1858, Der Jura: Tubingen, H. Laupp.

Reboul, Henri, 1833, Geologie de la periode quaternaire et introduction a l'histoire ancienne: Paris, Masson.

Rogers, H. D., 1858, The geology of Pennsylvania, a government survey, with a general view of the geology of the United States; essays on the coal formation and its fossils and description of the coal fields of North America and Great Britain: Philadelphia, Lippincott.

Schenck, H. G., and Muller, S. W., 1941, Stratigraphic terminology: Geol. Soc. Am. Bull., v. 52, p. 1419–1426.

Schimper, W. P., 1874, Traité de paléontologie végétale (v. 3): Paris, J. B. Bailliére et fils.

Sedgwick, Adam, 1831, Address, on announcing the first award of the Wollaston Prize: Geol. Soc. London Proc., v. 1, no. 20, p. 270–279.

———, 1838, A synopsis of the English series of stratified rocks inferior to the Old Red Sandstone—with an attempt to determine the successive natural groups and formations: Geol. Soc. London Proc., v. 2, no. 58, p. 675–685.

———, and Murchison, R. I., 1836, On the Silurian and Cambrian Systems, exhibiting the order in which the older sedimentary strata succeed each other in England and Wales: British Assoc. Adv. Sci. Report 5th meeting, August 1835, p. 59–61.

———, and Murchison, R. I., 1839, On the older rocks of Devonshire and Cornwall: Geol. Soc. London Proc., v. 3, no. 63, p. 121–123.

Steno, Nicolaus, 1669, De solido intra solidum naturaliter contento dissertationis prodromus: Florence.

Strachey, John, 1719, A curious description of the strata observ'd in the coal-mines of Mendip in Sommersetshire; being a letter of John Strachey Esq. to Dr. Robert Welsted, M. D. and R. S. Soc. and by him communicated to the society: Royal Soc. [London] Philos. Trans., v. 30, no. 360, p. 968–973.

———, 1725, An account of the strata in coal-mines, etc.: Royal Soc. [London] Philos. Trans., v. 33, no. 391, p. 395–398.

Umbgrove, J. H. F., 1938, Geological history of the East Indies: Am. Assoc. Petroleum Geologists Bull., v. 22, p. 1–70.

Werner, A. G., 1787, Kurze Klassification und beschreibung der verschieden Gebirgsarten: Dresden, Walther Hofbuchhandlung.

Williams, H. S., 1891, Correlation papers, Devonian and Carboniferous: U.S. Geol. Survey Bull. 80.

Williamson, W. C., 1834, On the distribution of organic remains in the Lias Series of Yorkshire with a view to facilitate its identification by giving the situation of its fossils: Geol. Soc. London Proc., v. 2, no. 36, p. 82–83.

———, 1836, On the distribution of organic remains in the Oolitic formations on the coast of Yorkshire: Geol. Soc. London Proc., v. 2, no. 47, p. 429–432.

———, 1838, On the distribution of organic remains in part of the Oolitic Series on the coast of Yorkshire; Geol. Soc. London Proc., v. 2, no. 57, p. 671–672.

Wilmarth, M. G., 1925, The geologic time classification of the United States Geological Survey compared with other classifications: U.S. Geol. Survey Bull. 769, 138 p.

Winchell, Alexander, 1870, On the geological age and equivalents of the Marshall group: Am. Philos. Soc. Proc., v. 11, p. 57–82, 385–418.

Ziegler, A. M., 1965, Silurian marine communities and their environmental significance: Nature, v. 207, no. 4994, p. 270–272.

Zittel, K. A., von, 1899, Geschichte der Geologie und Paläontologie: R. Oldenbourg, München und Leipzig, 868 p.

Index

Adam's birthdate, 20
Agricola, 3, 4
Alberti, Friedrich August von, 78–80
Alluvial rocks of Werner, 37
Alluvium of Buckland, 77
Altitudinal gradient, 139
American Association of Petroleum Geologists, 136
Arab scholars, 16
Arduino, Giovanni, 28, 39, 61, 62, 64, 65, 76, 108; stratigraphic conclusions, 33–34
Aristotle, 15
Arkell, W. J., 75
Artinsk Stage, 95
Aube area, France, 72
Aufgeschwemmte-Gebirge, 37, 77
Australia, 139
Auvergne area, France, 19, 40

Bacon, Roger, 1
Banks, Sir Joseph, 57
Barrande, Joachim, 96
Barrandium, 96
Bath Agricultural Society, 53
Bathymetric gradient, 139
Bauer, George, 3, 4
Beaumont, Elie de, 71, 75, 79
Beche, Sir Henry de la, 88
Belt Series, 120
Berry, W. B. N., 5, 134
Beyrich, Heinrich Ernst von, 112
Biochemistry, 141
Biogeographic provinces, 130–131
Biogeography, 141
Biological investigations in stratigraphic paleontology, 137–143

Black, James, 20
Black Jura, 75
Blasius, Professor, 92
Bordeaux basin, 109
Bottom types, 139
Boundaries of Oppel's zones, 128
Boundaries of time intervals, 137–139
Bracheux, sands of, 113
Brongniart, Alexandre, 61, 65, 68, 105
Bruno, Giordano, 17
Brown Jura, 75
Buch, Leopold von, 40, 75, 76, 92
Buckland, William, 12, 77, 81, 83; and Lyell, Charles, 104–105
Buffon, Comte de: concept of evolution, 42–45; earth history, 18–19
Bunter Sandstone, 32, 78, 79, 80
Burnet, Thomas, 18
Byzantine scholars, 16

California Tertiary succession, 135
Callomon, J. H., 140, 141
Cambrian, 63, 89, 95, 96, 97, 98, 116, 117, 138, 140; history of the growth of, 80–88
Caradoc Sandstone, 84
Carboniferous, 63, 88, 89, 90, 92, 93, 94, 117; history of the growth, 66–68; vs. Pennsylvanian and Mississippian, 99–102
Carboniferous Order of Conybeare and Phillips, 66–67, 83
Cary, John, 57
Catastrophism, 11, 12, 13, 52–53, 60

Cayugan, 63
Cenozoic, 116–118
Chalk, The, 69–72, 74
Chamberlain, T. C., and Salisbury, R. D., 72–73, 99, 101
Clerk, John, 20
Coal Measures, 53, 54, 66–68, 74, 88, 93, 100
Comanche Series, 72
Communities, 133
Comparative anatomy, 138
Congregations, diagnostic, 5, 128
Conybeare, W. D., and Phillips, William, 7, 8, 66, 67, 68, 69, 70, 71, 74, 76, 83
Copernicus, 2
Correlation between biogeographic provinces, 143
Correlation Papers, 101
Correlation web, 10, 138
Creation, Divine, 13, 34
Creation, The, 17
Cretaceous, 63, 76, 117; history of the growth, 69–73
Cuvier, Georges, 12, 57, 105; catastrophism, 43–44, 52–53, 124; evolution, 43–45; and Brongniart, 61, 65–66

Dark Ages, 16
Darwin, Charles, 46–49; and Wallace, 41, 42, 45, 50, 51
Da Vinci, Leonardo, 1, 16–17
Davy, Sir Humphrey, 81
Dean, W. T., 130, 131
Dechen, H. von, 78
Delmontian Stage, 135
Deluge, Noachian, 12, 17, 29, 34, 44, 77
Descartes, 2
Descriptive generalizations, 2
Deshayes, Gerard, 66, 106
Desmarest, Nicolas, 19, 39
Desnoyers, Jules, 106
Desnoyers, Paul G., 77
Desor, Pierre, 124
Devon and Cornwall, 81, 88–90
Devonian, 63, 68, 93, 97, 101, 117; history of the growth, 88–91

Diener, C., 80
Diluvium of Buckland, 77
Diodorus Sciculus, 3
DNA, 48
Donovan, D. T., 130, 131
Dufrenoy, O. J., 71, 75
Dumont, Andre, 91
Dutch East Indies, Tertiary of, 134, 136

Ecology, 141
Economic motive, 3, 4
Eifel area, Germany, 90–91
Embryology, 138
Emmons, Ebenezer, 96
Empedocles of Argigentum, 14; notion of evolution, 42
Environmental changes and reckoning time, 143
Eocene, 117; history of the growth, 107–114
Epochs of the Tertiary, 63, 121
Eras, 115–120
Evolution through natural selection, 9, 10, 26
Extinction of phyletic lineages, 132

Facies, 73
Faluns of the Loire, 109, 110
Faunal province, 139
Flötz-Gebürge, 29, 30, 31, 78
Flötz rocks, 66, 67, 69
Flötz-Schichten, 29, 31, 37
Forbes, Edward, 111
Formation, cartographic rock unit, 7, 61, 62, 63
Formation, eighteenth century German use, 8, 79
Fuchsel, George Christian, 32–33, 36, 78

Galileo, 2
Gang-Gebürge, 29, 31
Gebürge, Lehmann's understanding of, 30
Geike, Archibald, 99
Geinitz, H. B., and Gutbier, A., von, 95
Gemmellaro, G. G., 95

Index

Gene frequency, 48
Genesis, Book of, 17, 20, 60
Genetics, 138, 141
Geological Society of America, 136
Geological Society of London, 55, 58, 80, 81, 82, 84, 104, 116,
Giraud-Soulavie, Jean Louis Abbé, 50–51, 56, 57
Glacial deposits and the Quaternary, 77
Glaciation and the Pleistocene, 111–112
Gosselet, Jules, 91
Gothlandian, 99
Gould, S. J., 12, 13
Gradualism, 13
Grand Canyon Series, 120
Greenough, George Bellas, 58
Gressly, Armanz, 73
Group of formations, 8, 62
Guadalupe Mountains, Texas, 95
Guettard, Jean, 19
Gulf Series, 72
Gumbel, C. W. von, 80

Hall, James, 97
Halloy, J. J. d'Omalius d', 66, 67, 68, 70, 71
Harz Mountains, 31
Hausmann, F. L., 78
Hawn, Major Frederick, 95
Hellenic Greeks, 14–15
Heraclitus, 14
Herodotus, 14
Hill, R. T., 72
Historic time, 141
Hoffman, F., 78
Hooke, Robert, 18, 38, 51, 57
Howarth, M. K., 130, 131
Humboldt, Alexander von, 73, 74, 75, 92, 105
Hunter, Louis, 126
Hutton, James, 11, 12, 13, 28, 40, 60, 104; on uniformitarianism, 20–23

International Geological Congress, First, 7; Second, 7; Eighth, 7
Interpretive generalization, 2

Interpretive time scale, 63
Invasion of biogeographic province, 132
Isopach diagrams, 130

Jameson, Robert, 43
Jurassic, 53, 62, 79, 117, 141; D'Orbigny's subdivision, 122–125; history of the growth, 73–76; Oppel's subdivision, 126–129
Jura-Kalkstein, 73, 74
Jura Mountains, 71

Kainozoic, 116, 117
Karpinsky, A., 95
Keewatin Series, 120
Kepler, 2
Keuper Marls and Clays, 78, 79, 80
Keyserling, Count von, 92
King, W., 95
Kleinpell, R. M., 135–136
Koksharof, Lieutenant, 92
Koninck, Laurent G. de, 68

Laboratory studies of evolution, 141
Lamarck, Chevalier de, 43; concept of evolution, 44–45
Lapworth, Charles, 5, 87; on Ordovician, 97–99
La Roche, C., 78
Latitudinal gradient, 139
Lehmann, Johann, 33, 34, 36, 37, 61, 78; earth history, 28–31
Leupold, Wilhelm, and Vlerk, I. M. van der, 134
Lewis, Reverend T. T., 83
Leymerie, A., 72
Lias, 54, 74, 75, 79
Linnaean Society of London, 41, 46, 104
Lithofacies maps, 129–130
Llandeilo Flagstones, 84, 85
Llandovery (Early Silurian) paleogeography of Wales, 142–146
London Basin, 108–110
Lonsdale, William, 89–90, 92
Lower Silurian, 84
Lucretius, 15
Ludlow Rocks, 84, 85

Luisian Stage, 135
Lyceum, 15
Lyell, Charles, 6, 11, 12, 13, 28, 60, 117, 121, 122, 124, 137; influence on Darwin, 46; on the Tertiary Epochs, 66, 103–114; on uniformity, 22

Magnesian Limestone, 93, 95, 117
Mallory, V. S., 135
Malthus, T. R., 46
Mammals, in Tertiary correlation, 114
Mantell, Gideon A., 71
Marcou, Jules, 73, 124
Marker points, 10, 42, 137, 138, 140
Marsupials, 139
Mayer-Eymar, Carl, 125
McCoy, Frederick, 68, 87
Mediterranean Ammonite Province, 130, 131
Mendel, Gregor, 48
Merian, Peter, 78
Mesozoic, 116–118
Middle Ages, 11, 16
Miller, T. G., 140
Millstone Grit, 67, 68
Mining activity, influence of, 3–4, 28–29
Miocene, 6, 117; history of the growth, 107–114
Miocene stages in California, 135
Mississippian, 73; history of the growth, 99–102
Mohnian stage, 135
Mojsisovics, E., 80
Montbret, Baron de, 69
Montmollin, Auguste de, 71
Morlot, C. A. von, 77
Mountain Limestone, 67, 68
Muller, S. W., 7, 8
Murchison, R. I., 98, 99, 105, 121; on Devonian, 88–91; on Permian, 91–95; Russian excursions, 91–95; on Silurian, 80–88
Muschelkalk, 32, 74, 78, 79, 80

Natural laws, 14
Natural selection, 41, 42, 45–49

Neocatastrophism, 52–53
Neocomian Beds, 71
Neptunists, 39, 60
New Red Formation, 117
New York State, 63
New York State Geological Survey, 96
New York System, 97
Newer Pliocene, 107–111
Niagaran, 63
Nile River, 14
Northwest European Ammonite Province, 130, 131

Ochoan Stage, 63
Oeynhausen, C. von, 78
Old graywackes, 80, 84, 88
Old Red Sandstone, 66, 67, 68, 83, 84, 85, 87, 88, 89, 90, 92, 97, 98, 116, 117
Older Pliocene, 107–111
Oligocene, 112
Oolite, 54, 71, 74, 75, 117
Oppel, Albert, 10, 73, 75, 130, 133, 134, 135, 136, 137; principle, 126–129
Orbigny, Alcide d', 72, 122–125, 128
Ordovician, 63, 138; history of the growth, 95–99
Organic evolution, 9; Greek concept of, 42, 45
Oswegan, 63
Overlapping stratigraphic ranges, 127–128
Owen, David Dale, 100

Paleocene, 112–113
Paleozoic, 98, 116–118
Palaeozoic series of Sedgwick, 116
Palissy, Bernard, 17
Paris Basin, Tertiary of, 61, 65, 66, 108–110
Pennsylvania State Geological Survey, 100
Pennsylvanian, 73; history of growth, 99–102
Pennsylvanian Coal Measures, 100
Period as a unit in the time scale, 7–9

Index

Permian, 30, 33, 63, 79, 117; history of the growth, 91–95
Phase, 7, 8, 9, 128
Phillips, John, 68, 116, 117; on Devonian faunas, 90–91
Phyletic lineages, 132
Placental mammals, 139
Plato, 15
Playfair, John, 20, 21, 28, 60
Pleistocene, 111–112
Pliny the Elder, 15, 16
Pliocene, 117; history of growth, 106–114
Pompeii, 16
Precambrian, 115–116, 118–120
Primitive rocks, 34, 37, 61, 62, 65, 115
Primordial fauna of Barrande, 96; recognized in Norway and Sweden, 96
Prince Henry of Chambord, 96
Principle, 2–3
Principle of faunal (and floral) succession, 50, 51, 56, 57, 59, 61, 62, 63, 86, 103, 122
Principle of organic evolution through natural selection, 50, 59, 129, 137
Principle of uniformity in nature's processes, 11–12, 22–23, 28, 40, 50, 104
Productus Limestone, 95
Proterozoic, 120
Pythagoras, 14

Quaternary, 103; history of growth, 76–78
Quenstedt, F. A., 75, 125, 126

Reboul, Henri, P. I., 77
Recent, 107, 117
Relizian Stage, 135
Renaissance, 2, 16, 42
Rhaetic, 79
Rhenish Devonian faunas, 91
Rhineland travels of Murchison and Sedgwick, 90–91
Richardson, Reverend Benjamin, 55, 56

Roemer, Ferdinand, 91
Rogers, H. D., 100
Rome, scientific inquiry in, 15–17
Rotliegendes, 31, 94
Royal Society of Edinburgh, 20

Salisbury, R. D. and Chamberlain, T. C., 72–73, 99, 101
Salt Ranges, 95
Salter, John W., 87, 97
Sandberg, Guido and Fridolin von, 91
Saucesian Stage, 135
Saxony, Germany, A. Werner's geologic work in, 36, 69
Schenck, H. G., and Muller, S. W., 7, 8
Schimper, W. P., 5; on Paleocene, 112–113
Schlotheim, E. F. von, 78
Schwabian District of Jura Mountains, 126, 127
Secondary, 34, 61, 64, 65, 69, 81, 115
Sedgwick, Adams, 58, 91, 95, 97, 98, 116; on Cambrian, 80–88; on Devonian, 88–90
Series of formations, 8, 62, 63
Sezanne, travertines of, 113
Shumard, Benjamin Franklin, 95
Sicily: Lyell's travels, 105; Newer Pliocene, 107; Permian, 95
Silurian, 63, 89, 90, 92, 93, 95, 97, 98, 99, 116, 117, 138, 142, 143; history of the growth, 80–88
Silurian brachiopod communities in Wales, 143–146
Smith, William, 51, 62, 69, 73, 74, 76, 86, 122; geologic work, 53–59; list of fossil aggregates, 56; map of England and Wales, 55, 56, 57, 58
Soisson, lignites of, 113
Somerset, canal, 53, 54; stratigraphic succession in, 26–27, 56
Sowerby, George, 90
Species, 138
Stage: d'Orbigny, 122–125; Oppel, 128–129

Steensen, Niels, see Steno, Nicolaus
Steno, Nicolaus, 2, 24, 38, 40, 108; superposition, 24-25
Strabo, 15
Strachey, John, 26-27
Subapennine Hills, 107-109
Subcarboniferous, 100
Superga Hills, 108, 109
Superposition, 23-25, 50, 56

Taconic System, 96, 97
Teall, J. J. H., 99
Terrain, French use of, 8
Terrain Anthraxifere, 67
Terrain Bituminifere, 66
Terrain Creyace, 69, 70, 71, 72
Terrain Houiller, 67
Terrain Secondaire, 78
Tertiary, 34, 61, 62, 63, 76, 77, 115, 121; history of the growth, 64-66; Lyell's subdivisions, 103-114
Tertiary floras, 112-113
Thuringia, 32-33
Thurmann, Jules, 124
Time, 6
Time interval, beginning of, 10
Time "line," 140
Time "plane," 140
Time unit, 5, 6, 7, 9, 10
Time-stratigraphic series, 8, 63
Time-stratigraphic unit, 6, 7, 8
Timor Island, 95
Touraine Basin, 108-110
Townsend, Joseph, 55
Transition Rocks, 37, 39, 61, 62, 67, 115; Murchison and Sedgwick's work in Wales, 80-87
Triassic, 32, 53, 93, 117; history of the growth, 78-80
Tuscany, 107-109
Type areas, 6; supplemental, 6

Umbgrove, J. H. F., 134
Uniformitarianism, 13, 19, 60, 124
Uniformitarianism, methodological, 13
Uniformitarianism, substantive, 13
United States Geological Survey, 102, 136

Upper Cambrian Group, 85
Upper Carboniferous, 100
Ural Mountains, 92, 93
Ussher, Bishop, 12, 120

Valisnieri, Antonio, 64, 76
Verneuil, Edouard de, 91, 92
Vertical stratigraphic distribution of a species, 127
Vesuvius, eruption of, 16
Vienna Basin, 109
Vivarais, France, district of, 51-52
Voigt, Johann, 39
Volcanoes, 15, 16, 18, 19, 25, 40
Voltaire, 19
Vulcanists, 39, 60

Waagen, W., 80, 95
Wallace, Alfred Russel, 50-51; concept of evolution, 46; and Darwin, 41, 42, 45
Wanner, J., 95
Wealden Formation, 71
Wenlock Limestone, 84
Werner, Abraham G., 8, 12, 28, 34, 35, 60, 61, 62, 66, 67, 69, 74, 77, 78; concept of earth history, 36-40
White Chalk, 72
White Jura, 75
Williams, Henry Shaler, 101
Williamson, W. C., 126
Wilmarth, M. Grace, 120
Winchell, Alexander, 100
Wollaston Medal, 58-59
Wurttemberg, 125, 127

Xanthus the Lydian, 14
Xenophanes of Colophon, 14

Yorkshire Coast, 126

Zechstein, 31, 32, 79, 93, 94
Zemorrian Stage, 135
Zeigler, A. M., 143-146
Zone, Oppel, 128
Zone, time-stratigraphic, 7, 8
Zone names, 133
Zonengruppen, 128

d'Orbigny had described. Following their work, many more stage names were proposed by other workers of whom Carl Mayer-Eymar was perhaps the most prolific in establishing new stages by recognizing slightly different faunal aggregates from place to place. The naming of new stages became more a sport than a scientific endeavor as most of the stages put forth by all the workers who followed d'Orbigny have proven to be meaningless variations or divisions of d'Orbigny's original ones.

While a few workers, most of whom were French and Swiss, engaged in the game of naming new stages, other geologists found that they were unable to recognize d'Orbigny's divisions of the Jurassic in their parts of the world. In Germany, Friedrich Quenstedt, as he studied a sequence of Jurassic rocks in Württemberg very carefully and in minute detail, measuring thicknesses of strata exactly and noting the precise stratal position of each fossil, found that he could not use the d'Orbigny stages; his criticism of d'Orbigny's method centered around the acceptance, as the diagnostic faunal aggregate, of species from many strata in many localities, lumped together without enough regard for their precise stratigraphic ranges. He maintained that geologists should begin their attempts to subdivide the periods with a very detailed study of the stratigraphic ranges of species, that they should examine all strata, centimeter by centimeter, for fossils, and note precisely where each fossil was found in each stratigraphic section. Then, analysis of these details would lead, through comparison of the ranges of the species in many stratigraphic sections in many localities, to a fuller understanding of the succession of faunas. From such an understanding, Quenstedt asserted, the subdivision of the periods might proceed.

Quenstedt was not the only geologist of his time who was interested in determining the precise stratigraphic ranges of species through rock sequences although he championed the cause more vociferously than others. Painstaking studies of the sort he proclaimed were necessary had been and were being carried out in many places. Louis Hunter (1836) and W. C. Williamson (1834, 1836, 1838) examined Jurassic strata exposed along the Yorkshire coast of England, measuring the stratal succession with care and noting precisely the stratigraphic ranges of the several species of ammonites found there. Their studies are examples of the kind of precise stratigraphic work being carried out in the 1830's and 1840's in many parts of Europe.

Oppel's Principle

Although Quenstedt had declared that an overall picture of stratal and temporal relationships of fossiliferous rocks could be obtained by comparing stratigraphic and fossil range details from many stratigraphic sections from many different areas, he worked in only a relatively restricted area and did not derive any widely applicable principle for doing so. The method by which all fossiliferous strata could be divided into small-magnitude, clearly delimited units was put forth by one of Quenstedt's students at the University of Tübingen, Albert Oppel. Oppel studied Jurassic strata and the fossils found in them during the 1840's and 1850's. He followed in his teacher's footsteps by carefully studying local stratigraphic details and determining the stratigraphic ranges of many species as precisely as possible. He began his studies in the Schwabian district of the Jura Mountains. Then, after measuring many stratigraphic sections in that area, he traveled widely in Europe, examining collections of Jurassic fossils and stratal sequences. From this extensive background of data concerning the Jurassic deposits and fossils in Europe, Oppel (1856–1858) formulated the principle which permits establishment of the small-magnitude units of the prehistoric time scale and also paved the way toward establishing clearly delimited time units of any magnitude in that scale. His work gave real substance to the contributions paleontologists can make toward comprehension of the history of the earth.

Oppel began by determining the stratigraphic ranges of as many species of fossils as he could find in the Jurassic strata exposed in the Württemberg and Schwabian areas in Germany. He then extended his studies to include the Jurassic strata in France, Switzerland, and England. He was careful to note as precisely as possible the stratigraphic ranges of all the fossils found, particularly the ammonites, in every stratigraphic section studied. Oppel took care (1856–1858, p. 3) to "investigate the vertical distribution of each individual species at many different places ignoring the mineralogic character of the beds." He then plotted the stratigraphic range or vertical stratigraphic distribution of every species found in every stratigraphic section. From analysis of these plots, he quickly realized that there were groups of strata characterized by closely similar fossil aggregates. He believed these bodies of strata were time correlative. Then, as he studied the stratigraphic ranges of the several species closely, making a sort of composite series of stratigraphic sections so that as many species as possible might be included in his analysis, Oppel saw that the stratigraphic ranges of some species were short (being found with only one aggregate), others were very long, and most were of some intermediate length. Species appeared and disappeared in a random manner in the stratigraphic sections he had plotted. The different ranges of species led Oppel to realize that the distinctive faunal aggregates which characterize certain groups of strata were the result of the overlapping stratigraphic ranges of species. Using certain appearances of new species, Oppel discovered he could quite clearly delimit the boundaries between the distinctive aggregates, and could thus distinguish a succession of distinctive aggregates, each of which was bounded at its base by appearances of certain new species, and at its top by appearances of certain other new species. To the strata bearing the diagnostic aggregates, or congregations, of species Oppel gave the name *zones*.

Oppel (1856–1858, p. 4, 13, 15) pointed out that commonly only a few species—those appearing for the first time and those last seen in the zone—were members of the congregation. Many species ranged through two or more zones. Oppel specifically mentioned that it was the overlapping stratigraphic ranges of species that gave rise to the congregation which typified a zone. Oppel's principle, then, is that the analysis of the stratigraphic ranges of species leads to recognition of congregations of species that are unique in time. The boundaries between congregations are chosen at the first appearance of one or more members of the congregation, because these appearances are unique events in time. The time span corresponding to a zone, a phase, is the time that elapsed between the appearance or appearances chosen as the basal boundary of one zone and the appearance or appearances chosen as the basal boundary of the next succeeding zone. Oppel's principle of analysing the overlapping ranges of species thus permits recognition of clear-cut time units. No previous principle put forward for the analysis of fossils led to such definite time intervals.

Oppel expanded his discussion by grouping zones into stages (*Zonengruppen*). A stage, he maintained, was a larger unit than a zone as it was characterized by a larger magnitude fossil aggregate. Oppel's "stage" was thus slightly different from d'Orbigny's, based, as it was, on analysis of the overlapping ranges of species, which d'Orbigny did not use. D'Orbigny considered that each group of species had been specially created and had existed only for the time duration of its stage. Oppel knew that the stratigraphic ranges of species were variable, and used this knowledge to recognize the distinctive faunal aggregates upon which he based his stages.

Oppel actually divided the Jurassic into zones, and his zone